Effective Construction Project Delivery

Titus Ebenezer Kwofie · Clinton Aigbavboa · Wellington Thwala

Effective Construction Project Delivery

Improving Communication Performance in Non-Traditional Procurement Systems

Springer

Titus Ebenezer Kwofie
Department of Construction Management
and Quantity Surveying
University of Johannesburg
Johannesburg, South Africa

Department of Architecture
Kwame Nkrumah University of Science
and Technology
Kumasi, Ghana

Clinton Aigbavboa
Department of Construction Management
and Quantity Surveying
University of Johannesburg
Johannesburg, South Africa

Wellington Thwala
Department of Construction Management
and Quantity Surveying
University of Johannesburg
Johannesburg, South Africa

ISBN 978-3-030-49373-8 ISBN 978-3-030-49374-5 (eBook)
https://doi.org/10.1007/978-3-030-49374-5

This Springer imprint is published by the registered company Springer Nature Switzerland AG
The registered company address is: Gewerbestrasse 11, 6330 Cham, Switzerland

Preface

Construction project delivery process is information intensive and the imperative for effective communication is critical to improving effective project delivery and outcome across various procurement typologies. The construction project teams rely on various forms of communication which must suit the context, procurement model and unique role-specific task functions. The context, procurement model and role-specific task function often pose enormous challenges to effective communication, thereby affecting project delivery. Additionally, there is also lack of understanding of specific communication style and forms that can optimize project team interactions and information sharing across various procurement typologies. Currently, much of the literature overly focuses on traditional procurement systems. Non-traditional procurement typologies are seemingly characterized by unique attributes and features that limit the application of existing knowledge. Likewise, the urgent necessity of the construction industry towards a more collaborative working, integrated and cooperation among project teams in project delivery and rapid increase in the adoption of non-traditional procurement systems makes it imperative for a more theoretical and practical insight into improving communication in non-traditional procurement systems.

Effectiveness of team communication for the design and delivery of construction projects is becoming increasingly important in non-traditional procurements due to the growing technical and organizational complexity of construction projects. Construction project teams are generally multi-disciplinary in nature from organizations exhibiting different organizational cultures. They also exhibit different levels of understanding, opinions, skills and rates of adoption of the available communication styles and tools as well as preferences for specific means of communication. This situation heightens the uncertainties to expected communication performance across various procurement typologies. Effectiveness of project delivery process is influenced by two key aspects of communication which are the communication acts (technical and social based) of team members, their preferences for using specific communication media, access to easy to use tools, the competences of team to facilitate, stimulate and motivate effective communication and understanding of the requirements and performance of communication in a given context and specific procurement system.

There is current imperative to improve communication in non-traditional procurement as well as impact of client characteristics on performance in order to optimize the benefits of non-traditional procurement typologies in construction project delivery. However, while the later has witnessed extensive attention by way of research, less is known about improving communication performance in non-traditional procurement typologies. Though generally literature abounds on communication problems and challenges, communication performance and measures have primarily focused on traditional procurement systems. These do not attend to the task and social-based interaction and information sharing in non-traditional procurement typologies. Likewise, the obvious well-acknowledged complexities and dynamism of the project-based structure and professional and organizational culture of the construction industry do obliterate the effectiveness of existing central tenets of effective communication practice as well as its wholesale applicability to procurement models of unique attributes. The questions can communication be improved by adopting existing knowledge which are primed for conventional procurement models? Your answer and mine will certainly be no. In order to engender effective communication outcomes on non-traditional procurement, there is the need to develop and adopt approaches that are specific to it context and fit its characteristics. Hence, this book comes at a prime moment in the urge to generate understanding and knowledge that is suitable for effective communication in non-traditional procurement topologies. Indeed from evidence gathered from both theoretical and practical perspective, there is compelling evidence to affirm that non-traditional procurement typologies indeed experience communication problems and challenges that are unique compared to those experienced in conventional procurement systems. As already established in extant literature, most emerging management practices and procurement arrangements have failed to engender effective communication practices, interactions and information sharing that respond to the known structural and cultural conditions typical of the sector especially in non-traditional procurements. The need to generate knowledge and understanding can engender effective communication performance in non-traditional procurement typologies in compelling and critical to optimize its benefits.

In response, this book consisting four parts has the first offering an understanding of the value of communication to construction project delivery espousing the theoretical and conceptual position. Effective communication is deemed as a key tenet of delivery success by gaining insight into the key component of effective communication, information sharing and task and social-based interactions in non-traditional procurement typologies. In the second and third parts, an attempt is made to offer an in-depth understanding of the unique attributes of non-traditional procurement typologies and their implication for effective communication performance. This offers a foundation a clear approach to conceptualizing and measuring communication performance in non-traditional procurement typologies in part three. Given the current emerging attributes of the industry today with increasing era of digitalization, virtuality and behavioural uncertainties, gaining a practical knowledge on the way forward in communicating in this era is significant. This book recognizes the fact that, given that there is no bestspoke no single communication paradigm and strategies that fit all purposes and context but rather an urge for project team to adapt and operationalize strategies that suit the context and the given attribute of the procurement and management systems adopted for the project delivery. It is expected that by such approach, there can be high certainty of improved communication, interactions and information sharing that can engender the needed project delivery effectiveness.

Johannesburg, South Africa
April 2020

Titus Ebenezer Kwofie
Clinton Aigbavboa
Wellington Thwala

Acknowledgements We express our appreciation and befittingly acknowledge all those who made significant contribution to this book from vision to fruition. Special mention must be made to students of the Department of Construction Management and Quantity Surveying, University of Johannesburg, South Africa, for helping in the data collection and organization. We are also eternally grateful to the Head of Department, Construction Management and Quantity Surveying, University of Johannesburg, South Africa, and the Postgraduate School (PGS).

Contents

About the Authors

Titus Ebenezer Kwofie is a chartered architect by profession and a senior lecturer at the Kwame Nkrumah University of Science and Technology (KNUST) in Kumasi, Ghana. Before joining the Department of Architecture, he served as the Director of Works at the Koforidua Polytechnic (now Koforidua Technical University). He is currently a Postdoctoral Fellow at the Sustainable Human Settlement and Construction Research Centre and Department of Construction Management and Quantity Surveying at the University of Johannesburg in South Africa. He shares research interest in project team communication, team effectiveness, managerial efficiency and housing delivery. This has culminated into various research outputs that have been published in high impact journals and presentations at indexed conferences. He also serves as a regular reviewer for various Emerald, ASCE, Elsevier, Taylor & Francis and springer built environment journals. He has also been a secretary and deputy secretary to organizing committees of some international conferences.

Clinton Aigbavboa is a professor at the Department of Construction Management and Quantity Surveying, University of Johannesburg, South Africa. Before joining academia, he was involved as quantity surveyor on several infrastructural projects, both in Nigeria and South Africa. He is the immediate past Vice Dean of the Faculty of Engineering and Built Environment, University of Johannesburg, South Africa. He is a strong academic with extensive knowledge in practice, research, training and teaching. He is currently the Chair of Sustainable Human Settlement and Construction Research Centre at the University of Johannesburg. He is also an author of five research books that were published with Springer Nature and CRC Press. He is currently the editor of the Journal of Construction Project Management and Innovation (accredited by the DoHET) and has received national and international recognition in his field of research. The South Africa National Research Foundation rates him as a Young researcher with the potential of establishing himself within a five-year period.

Wellington Thwala is a professor at the Department of Construction Management and Quantity Surveying, University of Johannesburg, South Africa. He is the immediate past Head of the Department of Construction Management and Quantity Surveying, University of Johannesburg, South Africa. Currently, he is the Chair of SARChI in Sustainable Construction Management and Leadership in the Built Environment, FEBE, University of Johannesburg, South Africa. He offers research support and advice on construction-related issues to the construction industry in South Africa and government. He has extensive experience of providing consultancy for project leadership and management of construction projects and teaching project management subjects at the postgraduate level. He has an extensive industry experience with research focus on sustainable construction, leadership and project management. He is the editor-in-chief of International Journal of Construction Project Management and Innovation. He also serves as an editorial board member to various reputable international journals.

List of Figures

List of Tables

Part I
The Value of Communication: Theoretical Concepts and Context Nature of Communication

Chapter 1
General Introduction

Abstract This chapter offers an in-depth introduction into the rationale and concept behind this research book with the emphasis on the explanation of the key concept and theoretical basis of communication, the context and value of communication as well as an understanding of the human and technical aspects of communication in construction so as to place this book in proper perspective. A conscious attempt was made not to repeat what is already known about communication generally and specifically in the construction industry. However, the various attributes of communication and how it has been conceptualized in various contexts have also been highlighted and distinguished. The main objective of this book, which seeks to improve communication performance in non-traditional procurement models in construction project delivery, is also explained as well as an explanation of the scope and areas of concern that are addressed in each of the book chapters.

Keywords Value of communication · Construction communication · Non-traditional procurement

1.1 Introduction

Undoubtedly, gaining understanding of the significance of communication and conscious attempts to ensuring effective communication has longed been identified as a precursor to effective teamwork, organizational effectiveness and efficiency, process effectiveness and general improvement in performance and delivery success. Communication within construction context has been well acknowledged to be significant to success and managerial efficiency (Dainty et al. 2006; Kwofie et al. 2017; Emmitt and Gorse 2003). The construction industry in recent times has adopted numerous procurement models so as to improve process, delivery success and performance and optimize the benefits of these procurement models. In all these, Ling et al. (2015) and Manley and Chen (2016) have consistently outlined effective communication as a key factor through which the benefit of most procurement models can be optimized.

In recent times, non-traditional procurement models have gained credence and widely accepted as models that offer benefits above traditional procurement models

T. E. Kwofie et al., *Effective Construction Project Delivery*,
https://doi.org/10.1007/978-3-030-49374-5_1

(Manley and Chen 2016; Kwofie et al. 2019a, b; Song et al. 2009; Bresnen 2010; Chan et al. 2010). Dainty et al. (2006) shared the view of other authors and argued that communication has consistently remained a social activity and thus communication tasks and activities predominantly entail engaging in conversations, listening to colleagues, networking, collecting information, directing subordinates, writing letters or transferring information through electronic devices such as telephones or computers. However, while agreeing with this fundamental fact and truth about communication, it can further be extended that, even though these remain the core tenets of communication, forms expressed differ across various procurement typologies. Additionally, given the more digital nature information in the construction context has become, there is the need to examine and consider communication tenets to be more tailored to the particularities of specific procurement models (Kwofie et al. 2019b; El-Saboni et al. 2009). Against this background, this introductory chapter explores the nature and value of communication in non-traditional procurement systems and ways of what can be done towards engendering communication effectiveness to optimize the benefits of non-traditional procurement in the construction industry. Through this, the theoretical and contextual perspective of gaining an understanding towards effective communication in non-traditional procurement models in relation to the construction industry will be laid for the remaining chapters of this book.

1.2 Context and Value of Communication in Non-traditional Procurement and Way Forward

Construction procurement models and systems have generally become an increasingly diverse, complex with competing interests and task interrelated organization which heavily depend on effective communication for effectiveness, managerial efficiency, enhanced performance and success (Liu 2009; Gorse and Emmitt 2007; Xie 2002). However, in non-traditional procurement, the need for efficient and effective communication performance and behaviours has become a more increasingly an urgent necessity (Kwofie et al. 2019a, b; Xue et al. 2010). In the construction industry today, the gravitation towards increasing adoption of non-traditional procurement models is the notable fact that they are perceived as a business strategic models that offer enhanced performance and benefits beyond that often encountered in traditional procurement models (Manley and Chen 2016; Song et al. 2009; Bresnen 2010; Chan et al. 2010). For instance, collaborative procurement (working), alliancing, partnerships (e.g. PPP), relational contracting, strategic alliancing, integrated project management, joint venture, coalition, etc., have gained credence and been accepted globally as better procurement options to construction delivery over traditional methods (see Bresnen and Marshall 2000; Beach et al. 2005; Kling and Burley 1991; Abrahams and Cullen 1998; Xue et al. 2005; Xu et al. 2005; Hauck et al. 2004; Sillars and Kangari 2004; Anumba and Evbuomwan 1997, 1999; Ren et al. 2006; Kwok 1998; Pryke and Pearson 2006; Green et al. 2005; Walsh et al. 2004; Arbulu et al. 2003).

Davis and Walker (2009), Morwood et al. (2008), Love et al. (2010) and Kelly (2011) revealed that non-traditional procurement systems have potential attribute of evolving in response to changing market conditions and adapt to cultures that makes them more suitable for adequately managing complex infrastructure projects. In spite of these obvious facts and seemingly its being more preferred procurement option in the global construction industry, ensuring and enhancing effective communication outcomes and communication of conflict resolution strategy among stakeholders, participants, etc., have frequently been cited among the top critical success factors in non-traditional procurement (see Hauck et al. 2004; Chan et al. 2004; Xu et al. 2005; Yeung et al. 2007; Rowlinson 1999; McCabe 2001; Kumaraswamy et al. 2005; Diallo and Thuillier 2005). Kohli and Jensen (2010) recently affirmed that ineffective communication remains central to the emerging managerial ineffectiveness, conflict, lack of trust and ineffective collaborative working and practices in non-traditional procurement systems. This development only confirms the already existing knowledge of the significant role communication plays in the management and delivery of construction projects (see Cheung et al. 2004; Dainty et al. 2006; Gorse and Emmitt 2003; El-Saboni et al. 2009; Senescu et al. 2010).

Obviously, the unique attributes of non-traditional procurement models make the wholesale adoption of existing communication models, practices, behaviours, approaches that often suit traditional procurement models impractical. Hence, the need to pursue an agenda towards improving communication performance in non-traditional procurement has become an urgent reality. There have been few existing scholarships focusing on strategies and models to improve communication in construction project delivery (see Dainty et al. 2006; Gorse and Emmitt 2007; Liu 2009; Murray et al. 2000; Mead 1999; Xie et al. 2010). It could be admitted that these plethoras of scholarships provide insightful knowledge that are very relevant and still are towards improving communication in project delivery, the paucity in them is their lack of application across all procurement models and typologies. It is an open truth and well acknowledged among construction professionals and scholars that non-traditional procurement systems differ significantly in practice, models, structure, theories, approaches, etc., compared to the much accepted dominant traditional procurement model. Hence, the inability to apply these existing knowledge to improving communication in non-traditional procurement is an obvious theoretical and practical fact.

In the unique context of non-traditional procurement models, the value of effective communication towards individual task functions, teams, organizations and other stakeholders cannot be overlooked. In practical reality, effective communication of right information forms the bedrock of all actions, decisions, management and motivation in the delivery of the project (Manley and Chen 2016). In many countries, where non-traditional procurement models have been intensively used in the last two decades years as a means of more adequately managing complex infrastructure projects, existing scholarships suggest that many lack of effective communication has remained significant among other factors that has had large impact on time and cost outcomes, governance structures, supply chain relationships and commitment (Morwood et al. 2008; Love et al. 2010; Davis and Love 2011; Kelly 2011). Xue et al. (2010) and Deep et al. (2019) revealed the need to ensure effective communication as a single most important factor in project alliancing and collaborative working which are notable forms of non-traditional procurement. Manley and Chen (2016) also revealed that effective communication management and information control remains a significant mediating factor that influences the effect of client characteristics on the time and cost performance of collaborative infrastructure projects.

Blanc-Brude (2008) and Kwofie et al. (2019a) cited that communication and information imbalance as the immediate challenge that faces public–private partnerships (PPPs) in many developed and developing countries. A plausible reason ascribed to this development is the lack of thorough understanding of the unique contextual attributes of non-traditional procurement models. Hence, the need to generate knowledge and the dynamics of suiting communication approaches, management, practices to the dynamics and particularities of non-traditional models is paramount. This knowledge and understanding when generated will help garner and develop a mutually agreed communication *modus operandi* and robust and effective communication channels and media that suit the work and task functions and activities in non-traditional procurement systems and models. This can be deemed as the panacea for improving communication performance that is crucial towards managerial efficiencies, team effectiveness, delivery success, enhanced performance, reducing conflict and promoting trust and relationships as well as improving organizational processes in non-traditional procurement models. The importance of effective communication in the general construction industry and project delivery has extensively been outlined in many scholarships (see Dainty et al. 2006; Emmitt and Gorse 2007; Kwofie et al. 2016; Otter and Emmitt 2007, 2008; Ochieng and Price 2010; Diallo and Thuillier 2005). From these plethoras of literature, one can safely infer that communicating effectively is critical and forms a core facet of every aspect of project delivery process which must not be left to only the project manager but also every team participant and organizations on a given project. In the core attribute of this book, much emphasis will be given to the human and technical (digital) aspects of communication in construction. This is because understanding the human aspect communication is still relevant to the construction industry and the technical aspect thus embrace the digital nature of the fourth Industrial Revolution which has become the next new attribute of the construction industry as the digital age. Dainty et al. (2006) and Emmitt and Gorse (2007) have extensively given understanding to organizational, human aspects of communication and the principles of effective communication in the construction industry.

1.3 Understanding the Human and Technical (Digital) Aspects of Communication in Construction

As it has been profoundly acknowledged and extensively been elucidated, it is generally accepted that the general construction project context and environment presents more challenging field for the application of known communication practices proven to be effective in other sectors (Dainty et al. 2006; Otter and Emmitt 2007, 2008; Xie et al. 2010; Kwofie 2015). Generally, all project participants form a complex communication networks and channels which are intensive information dependent (Kwofie 2015; Otter and Emmitt 2007; Mead 1999; Xie et al. 2010).

Marshall-Ponting and Aouad (2005) revealed that the fundamental aspect leading to communication performance improvement is to focus on the human factors or the technical factors as the two distinct areas or both in the communication process. The focus on the human factors aspect of the communication process is key and thus should encompass the behaviours, skills, the project settings and attributes. The tools and technology adopted to aid communication and the communication context (environment) are perceived as the technical factors in the communication process (Yan 2009; Yang et al. 2007; Vanita and Yang 2006; Otter and Emmitt 2008; Otter 2005; Mead 1999). The emphasis on the technical aspect is premised on the theoretical and practical reality of the increased virtual nature of the construction process and large intensive nature of the information and data dependent on in a complex network and channels subjected to social interactions (Xie et al. 2010; Xie 2002).

The dimension of human aspect of communication in construction project context has extensively been dealt with by Dainty et al. (2006), Xie et al. (2010), Liu (2009) and Xie (2002). Given that the human aspect espoused by these studies is still relevant, thus the focus of this section shall be on the technical dimension. In recent times, it is well acknowledged that the construction project environment has become increasingly virtual, large information dependent, collaborative, digital in nature and process and ICT intensive (El-Saboni et al. 2009; Otter 2005; Otter and Emmitt 2007; Mead 1999).

Today, these attributes are changing the way the construction industry communicates, manages and stores information and interact on construction projects (El-Saboni et al. 2009; Otter 2005). With the current emerging attributes of the industry, a focus on the technical aspect of construction communication is appropriate because it holds the key to ameliorating the inherent challenges of virtuality, management and storage of information and bridging the digital divide so as to improve interactions and transactions individuals, teams and organizations in construction. This will be a sure way to improve the access, understanding, completeness of project information and errors in human interface and enhance working together in an integrated and collaborative way.

It must further be noted that high dependence on effective communication is not only peculiar to the construction industry or construction project delivery process. Even though the construction industry is perceived as an established sector and information intensive, further refinement to processes and protocols that suits the particularities of teams, organization, virtuality, procurement systems to facilitate communication is crucial as a sure way of eliminating any elements of uncertainties that may exist and is potential to affect the communication channels and networks necessary for project success. The industry has witnessed a wave of enthusiasm for the introduction of ICT tools, virtual and digital platforms for communication and information sharing. However, their effectiveness may be hampered by the uniqueness and particularities of project context and procurement models (Ibrahim et al. 2019; El-Saboni et al. 2009; Otter 2005; Otter and Emmitt 2007). In acknowledging this milieu and the centrality of effective communication in construction project delivery, the focus of the industry has been on development on more tools to enhance information sharing and communication (El-Saboni et al. 2009; Otter 2005; Otter and Emmitt 2007; Mead 1999).

It must be well noted that it is difficult to fit the human network of inter-organizational and interpersonal and technical aspect of communication to generic procurement systems and construction project context and process. Hence, the way forward is the move away from generic models for construction communication to and integration of human and technical aspect of construction communication to fit procurement systems.

1.4 The Objective and Focus of This Book

There exist various divergent views and argument concerning the way forward to improving construction communication especially in the wake of increasing virtuality, high volume of project information, urgent need for a more integrated and collaborative process, existence of information asymmetries and unique particularities of various procurement models (Ibrahim et al. 2019; Kwofie et al. 2015, 2019a; El-Saboni et al. 2009; Dainty et al. 2006; Emmitt and Gorse 2007; Otter 2005; Marshall-Ponting and Aouad 2005; Otter and Emmitt 2007; Liu 2009; Xie 2002). From mainstream project management genre and materials above enumerating these gaps, there is yet to be actual any scholarship focusing on communication performance in non-traditional procurement models and systems. Given that this gap is an imminent necessity, the focus of this book is extended and expands this area to include the virtual and digital dimensions of communication integrating the human and technical dimensions. This book perceives communication as not just a mere exchange of information between project teams but as a task function, action and activities that can motivate the understanding, access, protocols in information documentation, sharing, management, etc., in construction project delivery especially in non-traditional procurement models. It is expected that this book will offer the needed knowledge and understanding to engender effective communication in non-traditional procurement systems.

1.5 Summary

This chapter has primarily espoused the rationale and idea behind the conception of this research book giving the key explanation for the focus of the book relating to communication performance in non-traditional procurement systems and models. It has made a strong case for the need to align communication performance improvement and its practices and approaches to the procurement models rather than generic approach. It has also explained the key issues relating to human and technical dimensions of communication in construction and advocate for the need to integrate both in a quest to bring about improvement in communication effectiveness on construction project to engender the needed integration and collaboration given the more virtual and digital nature of the industry now.

The objective of the book, which is to explore the key tenets to improvement in communication performance in non-traditional procurement in construction project delivery, was also explained while detailing the scope and areas of concern that are contained in each of the book chapters. The theoretical and practical concepts in human and technical dimensions of construction communication are further explained in Chap. 2.

References

Abrahams, A., & Cullen, A. (1998). Project alliances in the construction industry. *Australian Construction Law Newsletter, 62*, 31–36.

Anumba, C. J., & Evbuomwan, N. F. O. (1997). Concurrent engineering in design-build projects. *Construction Management and Economics, 15*(3), 271–281.

Anumba, C. J., & Evbuomwan, N. F. O. (1999). A taxonomy for communication facets in concurrent life-cycle design and construction. *Computer-Aided Civil and Infrastructure Engineering, 14*(1), 37–44.

Arbulu, R. J., Tommelein, I. D., Walsh, K. D., & Hershauer, J. C. (2003). Value stream analysis of a re-engineered construction supply chain. *Building Research and Information, 31*(2), 161–171.

Beach, R., Webster, M., & Campbell, K. M. (2005). An evaluation of partnership development in the construction industry. *International Journal of Project Management, 23*(8), 611–621.

Blanc-Brude, F. (2008). *Public-private risk transfer—Public-private partnerships, long-term contracts and price discovery*. London: King's College.

Bresnen, M. (2010). Keeping it real? Constituting partnering through boundary objects. *Construction Management & Economics, 28*(6), 615–628.

Bresnen, M., & Marshall, N. (2000). Partnering in construction: A critical review of issues, problems and dilemmas. *Construction Management and Economics, 18*(2), 229–237.

Chan, A. P. C., Chan, D. W. M., Chiang, Y. H., Tang, B. S., Chan, E. H. W., & Ho, K. S. K. (2004). Exploring critical success factors for partnering in construction projects. *Journal of Construction Engineering and Management, 130*(2), 188–198.

Chan, A. P. C., Chan, D. W., & Yeung, J. F. (2010). *Relational contracting for construction excellence: Principles, practices and case studies*. Abingdon: Spon Press.

Chen, L., & Manley, K. (2014). Validation of an instrument to measure governance and performance on collaborative infrastructure projects. *Journal of Construction Engineering and Management, 140*(5).

Cheung, S. O., Suen, C. H., & Cheung, K. W. (2004). PPMS: A web-based construction project performance monitoring system. *Automation in Construction, 13,* 361–376.

Dainty, A. Moore, D., & Murray, M. (2006). *Communication in construction-theory and practice.* Abingdon, Oxon: Taylor & Francis.

Davis, P., & Love, P. (2011). Alliance contracting: Adding value through relationship development. *Engineering, Construction and Architectural Management, 18*(5), 444–461.

Davis, P. R., & Walker, D. H. T. (2009). Building capability in construction projects: A relationship-based approach. *Engineering, Construction and Architectural Management, 16*(5), 475–489.

Deep, S., Gajendran, T., & Jefferies, M. (2019). A systematic review of 'enablers of collaboration' among the participants in construction projects. *International Journal of Construction Management*, 1–13. https://doi.org/10.1080/15623599.2019.1596624.

Diallo, A., & Thuillier, D. (2005). The success of international development projects: Trust and communication: An African perspective. *International Journal of Project Management, 23*(3), 237–252.

El-Saboni, M., Aouad G., & Sabouni, A. (2009). Electronic communication systems effects on the success of construction projects in United Arab Emirates. *Advanced Engineering Informatics, 23*(1), 130–138.

Emmitt, S., & Gorse, C. A. (2003). *Construction Communication.* Oxford: Blackwell Publishing.

Emmitt, S., & Gorse, C. A. (2007). *Communication in Construction Teams.* Taylor & Francis, London.

Gorse, C. A., & Emmitt, S. (2003). Investigating interpersonal communication during the construction progress meetings: Challenges and opportunities. *Engineering, Construction and Architectural Management, 10*(4), 234–244.

Gorse, C. A., & Emmitt, S. (2007). Communication behaviour during management and design team meetings: A comparison of group interaction. *Construction Management and Economics, 25,* 1197–1213.

Green, S. D., Fernie, S., & Weller, S. (2005). Making sense of supply chain management: A comparative study of aerospace and construction. *Construction Management and Economics, 23*(6), 579–593.

Hauck, A. J., Walker, D. H. T., Hampson, K. D., & Peters, R. J. (2004). Project alliancing at national museum of Australia—Collaborative process. *Journal of Construction Engineering and Management, 130*(1), 143–152.

Ibrahim, C. K. I. C., Sabri, N. A. M., Belayutham, S., & Mahamadu, A. (2019). Exploring behavioural factors for information sharing in BIM projects in the Malaysian construction industry. *Built Environment Project and Asset Management, 9*(1), 15–28.

Kelly, J. (2011). *Cracking the VFM code: How to identify & deliver genuine value for money in collaborative contracting.* Sydney: Big Fig Publishing Limited, under the imprint of Intelligentsia Press.

Kling, E., & Burley, S. (1991). European community joint ventures, Chap. 1. In *Joint ventures in Europe* (p. 1). London: Butterworths.

Kohli, A., & Jensen, J. (2010). Assessing effective of supply chain collaboration. *Supply Chain Forum: An International Journal*, 2–16.

Kumaraswamy, M. M., Rahman, M. M., Ling, F. Y. Y., & Phng, S. T. (2005). Reconstructing cultures for relational contracting. *Journal of Construction Engineering and Management, 131*(10), 1065–1075.

Kwofie, E. (2015). *Contribution of unique features of mass housing projects to project team communication performance* (Unpublished Ph.D. thesis). Department of Building Technology, Kwame Nkrumah University of Science and Technology, Kumasi.

Kwofie, T. E., Adinyira, E., & Fugar, F. (2016). Modelling the effect of housing design unit contract packaging on mass housing project team communication performance. *Journal of Construction in Developing Countries, 21*(1), 55–70.

Kwofie, T. E., Adinyira, E., & Fugar, F. (2017). Theoretical and practical implications for engendering project team communication effectiveness in mass housing project delivery in Ghana. *Journal of Engineering, Design and Technology, 15*(6), 826–844.

Kwofie, T. E., Aigbavboa, C. O., & Machethe, S. O. (2019a). Nature of communication performance in non-traditional procurements in South Africa. *Engineering, Construction and Architectural Management, 26*(10), 2264–2288.

Kwofie, T. E., Aigbavboa, C. O., & Thwala, W. D. (2019b). Communication performance challenges in PPP projects: Cases of Ghana and South Africa. *Built Environment Project and Asset Management, 9*(5), 628–641.

Kwofie, T. E., Fugar, F., & Adinyira, E. (2015). Contribution of multiple construction site management features to project team communication effectiveness: The case of mass housing projects. *Engineering Project Organization Journal, 5,* 180–193.

Kwok, T. (1998). *Strategic alliances in construction: A study of contracting relationships and competitive advantage in public sector building works* (Ph.D. dissertation). Queensland University of Technology, Brisbane, Australia.

Ling, F., Peng Chong, T., Yan, N., Albert, T., & Asanga, G. (2015). Effect of adoption of relational contracting practices on relationship quality in public projects in Singapore null. *Engineering, Construction and Architectural Management, 22*(2), 169–189.

Liu, Y. (2009). *Critical factors for managing project team communication at the construction stage* (Ph.D. thesis). Submitted to the Polytechnic University of Hong Kong.

Love, P. E. D., Mistry, D., & Davis, P. R. (2010). Price competitive alliance projects: Identification of success factors for public clients. *Journal of Construction Engineering and Management, 136*(9), 947–956.

Manley, K., & Chen, L. (2016). The impact of client characteristics on the time and cost performance of collaborative infrastructure projects. *Engineering, Construction and Architectural Management, 23*(4), 511–532.

Marshall-Ponting, A. J., & Aouad, G. (2005). An nD modelling approach to improve communication process for construction. *Automation in Construction, 14*(3), 311–321.

McCabe, S. (2001). *Benchmarking in construction.* Oxford, UK: Blackwell Science.

Mead, S. T. (1999). *Communication effectiveness in intranet based construction projects* (Unpublished Ph.D. thesis). Department of Civil and Building Engineering, Loughborough University, Loughborough, UK.

Morwood, R., Scott, D., & Pitcher, I. (2008). *Alliancing a participant's guide: Real life experiences for constructors, designers, facilitators and clients.* AECOM.

Murray, M. D., Tookey, J. E., Langford, D. A., & Hardcastle, C. (2000). Project communication variables: A comparative study of US and UK construction industry perceptions. In A. Akintoye (Ed.), *16th Annual ARCOM Conference* (Vol. 2, pp. 813–822), September 6–8, 2000, Glasgow Caledonian University. Association of Researchers in Construction Management.

Ochieng, E. G., & Price, A. D. (2010). Managing cross-cultural communication in multi-cultural construction project teams: The case of Kenya and UK. *International Journal of Project Management, 28*(5), 449–460.

Otter, A., & Emmitt, S. (2007). Exploring effectiveness of team communication balancing synchronous and asynchronous communication in design teams. *Engineering, Construction and Architectural Management, 14*(5), 408–419.

Otter, A. F. H. J. (2005). *Design team communication using a project website* (Ph.D. thesis). Bouwstenen 98, Technische Universiteit Eindhoven, Eindhoven.

Otter, A., & Emmitt, S. (2008). Design team communication and design task complexity: The preference for dialogues. *Architectural, Engineering and Design Management, 4*(2), 121–129.

Pryke, S., & Pearson, S. (2006). Project governance: Case studies on financial incentives. *Building Research and Information, 34*(6), 534–545.

Ren, Z., Anumba, C. J., Hassan, T. M., Augenbroe, G., & Mangini, M. (2006). Collaborative project planning: A novel approach through an e-engineering hub—A case study of seismic risk analysis. *Computers in Industry, 57*(3), 218–230.

Rowlinson, S. (1999). Selection criteria. In S. Rowlinson & P. McDermott (Eds.), *Procurement systems: A guide to best practice* (pp. 276–299). London: E & FN Spon.
Senescu, R. R., Haymaker, J. R., & Anderson, D. J. (2010). PIP: A process communication web tool. In *Modelling and management of engineering processes*. Cambridge, UK: Cambridge University.
Sillars, D. N., & Kangari, R. (2004). Predicting organizational success within a project-based joint venture alliance. *Journal of Construction Engineering and Management, 130*(4), 500–508.
Song, L., Mohamed, Y., & AbouRizk, S. M. (2009). Early contractor involvement in design and its impact on construction schedule performance. *Journal of Management in Engineering, 25*(1), 12–20.
Vanita, A., & Yang, J. (2006). Communication protocol for building project management—The potential of I.T. enhanced approaches for the Indian building practice. In *Proceedings of the CIB W89 International Conference on Building Education and Research*, April 10–13, 2006, China, Hong Kong.
Walsh, K. D., Hershauer, J. C., Tommelein, I. D., & Walsh, T. A. (2004). Strategic positioning of inventory to match demand in a capital projects supply chain. *Journal of Construction Engineering and Management, 130*(6), 818–826.
Xie, C., Wu, D., Luo, J., & Hu, X. (2010). A case study of multi-team communications in construction design under supply chain partnering. *Supply Chain Management: An International Journal, 15*(5), 363–370.
Xie, X. (2002). *Communications in construction design* (Ph.D. thesis). Loughborough University, Loughborough.
Xu, T., Smith, N. J., & Bower, D. A. (2005). Forms of collaboration and project delivery in Chinese construction markets: Probable emergence of strategic alliances and design/build. *Journal of Management and Engineering, 21*(3), 100–109.
Xue, X., Shen, Q., & Ren, Z. (2010). Critical review of collaborative working in construction projects: Business environment and human behaviors. *Journal of Management in Engineering, 26*(4), 196–208.
Xue, X. L., Li, X. D., Shen, Q. P., & Wang, Y. W. (2005). An agent-based framework for supply chain coordination in construction. *Automation in Construction, 14*(3), 413–430.
Yan, L. (2009). *Critical factors for managing project communication among participants at the construction stage*. Published theses at Hong Kong Polytechnic University.
Yang, J., Ahuja, V., & Shankar, R. (2007). *Managing building projects through enchanted communication—An ICT based strategy for small and medium enterprises*. Paper presented at the CIB World Building Congress (pp. 2344–2357), Cape Town, South Africa, May, 2007.
Yeung, J. F. Y., Chan, A. P. C., & Chan, D. W. M. (2007). The definition of alliancing in construction as a Wittgenstein family resemblance concept. *International Journal of Project Management, 25*(3), 219–231.

Chapter 2
Understanding Communication in Construction Project Delivery

Abstract The concept of communication has become important factor to construction project delivery. Uncontestably, communication has been conceptualized in varied forms across various domains such as psychology, marketing, sociology and mass communication. This chapter provides a theoretical understanding of the concept of communication in construction context and provides its evolution in the industry. Owing to the varying understanding and conceptualization and different levels of perceptions of construction communication, the contextual definition is provided to position the concept in the construction domain as well as delineating the critical elements in construction communication process. As significance as communication is to construction project delivery, there are still certain aspects such as context and dynamics of communication that require and in-depth understanding to aid in efforts aimed at improving construction communication. This understanding is thus crucial to delineating both direct and indirect strategies that may help project teams to communicate effectively and ultimately improve the project performance and management.

Keywords Concept of communication · Project delivery · Dynamics of communication

2.1 Introduction

Guevara and Boyer (1981) posited that communication systems have remained the central nervous systems which make it possible for varied participants to undertake numerous tasks in an integrated and orderly manner and to coordinate their efforts and skills towards a common goal. However, from a theoretical perspective, the term communication has been conceptualized and defined in varied manner in various domains such as business, sociology, anthropology, psychology and most recently in the field of organizational communication (see Dainty et al. 2006; Salleh 2008; Adler et al. 2004; Morreale et al. 2006; Spitzberg and Chagnon 2009; Kwofie 2015). This gives the term a diverse background which is often though extensive but quite fragmented (Guevara and Boyer 1981; Dainty et al. 2006; Salleh 2008). Even in the construction project context, the term has also assumed varied conceptual definition,

T. E. Kwofie et al., *Effective Construction Project Delivery*,
https://doi.org/10.1007/978-3-030-49374-5_2

understanding and theoretical accentuation (see Thomas et al. 1998, 1999; Dainty et al. 2006; Sonnenwald 1996; Emmitt and Gorse 2007; Hoezen et al. 2010; Liu 2009; Xie et al. 2010). In this chapter, an understanding of the concept of communication as well as the main elements and attributes of the communication process are theoretically placed in perspective.

2.2 Theoretical Concepts in Communication

From the universal law of communication theory, it has been posited that communication forms part of all living beings through the use of sound, speech, visible changes, body movements, gestures in the best possible way to make others aware and understand their thoughts, feelings, problems, happiness or any other information (Jablin and Sias 2001; Goulden 1992). However, several and varied theories have been used to express conceptualized communication in various contexts. There seems to be a general consensus among most scholars of communication in support basically conceptualizing and perceiving communication basically as the process of transferring all forms of information from the sender to the recipient where the recipient decodes the information and acts accordingly (see Shannon and Weaver 1949; Schermerhorn et al. 1994; Richmond and McCroskey 1992; Te'eni et al. 2001; Jablin and Sias 2001; Monge and Eisenberg 1987; Salleh 2008; Weiner 2006; Dainty et al. 2006; Gorse and Emmitt 2007).

In the construction industry context, various theories such as the positional theory, attribution theory, relational theory, expectancy theory have all been used to conceptualize communication (Jablin and Sias 2001; Xie 2002; Xie et al. 2010; Liu 2009; Mead 1999; Kwofie et al. 2014; Kwofie 2015; Kwofie et al. 2016). The positional theory analyses organizational and team communication in terms of positional, relational and cultural roles performed and perceives network structure as formal and static (Jablin and Sias 2001; Monge et al. 1998).

The relational theory on the other hand relates behavioural patterns of individuals in developing and maintaining communication linkages within organizations and observing the interactions of people within a group and assumes that individuals often communicate in informal ways that are beyond their positions (Salleh 2008; Wilson and Sabee 2003; Spitzberg and Cupach 2002). In the case of the attribution theory, it offers an accurate explanation for the communication performance outcome in any context (Weiner 2006; Kwofie et al. 2016). Using this, Weiner (2006) revealed that external and internal factors are the main causal factors for the communication performance outcome in any context. The expectancy theory explains that communication must always situationally and culturally relevant expectations (Wilson and Sabee 2003; Jablin et al. 1994; Jablin and Sias 2001; Spitzberg and Cupach 2002). From most theories as applied to general communication and construction communication specifically, it can be deduced that cognitive and social dimensions are the two main tenets towards effective communication performance (Salleh 2008; Wilson and Sabee 2003). In the cognitive dimension, the emphasis is on the need to have requisite knowledge for message processing and the ability to process the knowledge for message production (Spitzberg and Cupach 2002; Wilson and Sabee 2003; Henderson 2004, 2008; Salleh 2008). The social dimension on the other hand explains the interactional aspect of the communication process (Spitzberg and Cupach 2002; Wilson and Sabee 2003; Henderson 2004).

From a theoretical position, it can be argued that construction project delivery process and construction context are information intensive and are an interaction process involving multidisciplinary teams and organizations (Sonnenwald 1996; Thomas et al. 1998, 1999). Recent research on communication in construction has extended communication tasks and function beyond just the composition of project-related information, messages, ideas to include a focus on the technical aspect such as the application of new technologies to facilitate distributed team and organizational interactions, collaboration to achieve effective coordination and communication between colocated and distributed project teams (Perry and Sanderson 1998; Wikforss and Lofgren 2007; El-Saboni et al. 2009). Hence, it must be accepted that the technical dimension of communication cannot be overlooked from both theoretical and conceptual perspective.

2.3 Communication as a Vital Tool in Construction Project Delivery

In the construction context, an all embracing definition capturing the reliance of communication to the industry refers to the 'timely and appropriate generation, collection, dissemination, storage and ultimate implementation of project information, ideas, goals, knowledge, etc., resulting in shared meaning and understanding for enhancing project delivery' (PMI 2008). This definition is an integration of those espoused by Zhao (2003), Brown (2001), Anumba and Evbuomwan (1999), Guevara and Boyer (1981), Sonnenwald (1996) and Pietroforte (1997). From this definition, it can be affirmed that communication in construction assumes a management approach and dimension, and such asserts its significance to the construction industry and project delivery process.

Construction processes are generally information intensive requiring large volume of information and documents to be generated and shared for the purposes of completing several process and tasks simultaneously (Cui et al. 2018; Kwofie et al. 2019). Pietroforte (1997) revealed that communication is critical to achieving and accomplishing common goals in construction building process. It is also recounted that in a typical construction environment and teamwork, communication can be a valuable channel of influence in changing personal, team and organizational relationships (Pietroforte 1997; Popple and Towndrow 1994). Sonnenwald (1996) posited that, without communication, it is almost near impossible to achieve effective integration and collaboration in construction project delivery.

Anumba and Evbuomwan (1999) agreed with Guevara and Boyer (1981) that especially in the construction industry, communication forms the central nervous systems affirm the possibility of dozens of team and organizational participants to undertake tasks in a integrated and orderly manner as well as coordinating their efforts and skills towards achieving a common goal. Gorse and Emmitt (2007) also affirmed that tasks and relational interactions among construction project teams are possible with effective communication. The design and construction process have overly been described as being a collaborative social act that heavily depend on effective interactions for task-based activities and social motives using a variety of communication tools and media (Gorse and Emmitt 2007; Dainty et al. 2006).

Gorse and Emmitt (2007) intimated that, generally, construction project teams and organizations in managing projects in their bid to achieve goals and discuss issues must of a necessity share task-base information through the exchange of ideas, information, opinions, knowledge and suggestions. Likewise, communication in construction is a necessity for forming and maintaining relationships, showing support, establishing values and resolving differences through interactions that have social and emotional content (Gorse and Emmitt 2007; Dainty et al. 2006; Otter 2005). Emmitt and Gorse (2003) revealed that intrapersonal, interpersonal, group, multi-group and mass communication are typologies of communication vital to construction industry and project delivery purposes.

Kreps (1989) alluded that intrapersonal communication is vital for individuals to process information, whereas interpersonal communication is a necessity for establishing and maintaining relationships. Group communication also helps in coordinating efforts and activities of work groups (Kreps 1989). From this, there is no doubt about the centrality and relevance of communication to construction tasks, team functions and relationship and organizational atmosphere for project success and team performance. It can be asserted that communication still remains the "life-blood" of construction building process and effectiveness of the management of the project.

2.4 Critical Elements in Construction Communication Process

Several scholarships have delineated the key elements in general communication, while others have been specific to certain domain and contexts (see Liu 2009; Sigband and Bell 1989; Shannon and Weaver 1949; Emmitt and Gorse 2003; Pietroforte 1997; Popple and Towndrow 1994; Sonnenwald 1996; Thomas et al. 1998; Guevara and Boyer 1981; Dainty et al. 2006). The relevance of these elements to construction communication is not in doubt. Hence, in an effort not to repeat same key elements of communication, an attempt is made to relate the relevant elements to the emerging context of the construction industry in respect of virtuality and digital divide so as to offer uniqueness and relevance to the discourse across the various project life cycle phases.

Many existing literature have predominantly identified communication process model, the communicators, information and message, communication channels and networks, and communication media as the key elements in communication especially in human communication (see Liu 2009; Sigband and Bell 1989; Shannon and Weaver 1949; Emmitt and Gorse 2003; Xie 2002; Pietroforte 1997; Popple and Towndrow 1994; Sonnenwald 1996; Thomas et al. 1998; Guevara and Boyer 1981; Dainty et al. 2006; Howard et al. 1998; Mead 1999; Gorse et al. 1999). Even though these elements have extensively been discussed in relation to general human communication, some notable discourse have been made with reference to construction project environment (see Dainty et al. 2006; Liu 2009; Xie 2002; Xie et al. 2010; Emmitt and Gorse 2003).

Mead (1999) on the other hand revealed that, in respect of ensuring effective communication in construction, roles, information, linkages (networks), barriers and communication metrics are five critical components that must be considered. From the communication model perspective, communication has predominantly been perceived as a linear process, and this is often in human communication practised in psychology and sociology (Schermerhorn et al. 1994; Shannon and Weaver 1949). However, it is well acknowledged that construction communication is more dynamic and complex limiting the application of such models (Thomas 1996; Thomas et al. 1998; Emmitt and Gorse 2003, 2007). In the communicators (sender and receiver) as a key element of communication, communication is understood as a human (sender) conceives (encodes) and becomes the source of the information and sent to a human (receiver) through an acceptable media who receive and decode the information (message) and may give feedback or not after receiving the information (Du 2004; Thomas et al. 1998). In a typical construction communication context, the sender uses both verbal and non-verbal elements conceiving the information and must ensure that all aimed receivers can be accessed and receive the information (Emmitt and Gorse 2003; Thomas et al. 1998; Dainty et al. 2006). In this dimension, Kwofie et al. (2016) and however, it must be admitted that, in this current era of the industry, information has taken a digital dimension, and thus, varied forms of approaches are being used to conceive digital information in virtual environment to enhance access and understanding of project information in communication.

According to Dainty et al. (2006) and Emmitt and Gorse (2007), information and message still remain a key element towards effective communication performance in construction. In this regard, it is said that the information or the message refers to the encoded idea, knowledge or meaning being transmitted from the sender to the receiver which can take the form of static documents or dynamic processed data (Dainty et al. 2006). Static information does not change, whereas dynamic information is always evolving and could be subjected to varying interpretations by receivers (Xie et al. 2010). In construction context, it is often suggested that communicated project-related information or message may take the form of documents, note, memo, letters, reports, sketches, drawing, graphical materials and physical models (Liu 2009). In recent times, 3D models in virtual environment and augmented reality have also been form of information in project environment assessed and utilized in the construction process (Otter 2005; Cheng and Teizer 2013).

Communication channels and networks have also remained key element of construction communication. They can be the formal or informal conduits through which the messages of information encoded flow from sender to receiver (Thomas et al. 1998; Xie 2002; Xie et al. 2010). In construction project-based context, formal channels are often regarded as official which does inform participants and deliver instructions including contract, change orders, schedules, policies, etc. (Thomas et al. 1998). Xie (2002) intimated that informal channels are also valuable in construction communication in the sense that in some cases they address the inadequacies of the formal channels even though some consider them as being inefficient (Sigband and Bell 1989; Hunter 1993).

Communication networks refer to the formal and informal linkages between project participants who access project-related information from a sender and receiver model (Dainty et al. 2006; Emmitt and Gorse 2007). It is often accepted that construction communication leans more towards organizational communication which is considered to form more complex networks and linkages than human communication (Eisenberg et al. 1985; Hunter 1993). This complexity stems from the theoretical foundation that construction project process involves large participants in dynamic roles and performs various tasks which are often changing from inception to closure. According to Emmitt and Gorse (2003), chain, wheel and comcon are the three basic communication models, and thus, an interplay of combination of these models is usually experienced in construction building process. Mead (1999) revealed the efficiency of communication networks and linkages are influenced by the nature of the context (environment), the type and nature of information and messages transmitted and the procurement models adopted. Overall, it must be well noted that the relations communicators in a communication network is more complex than how they are normally perceived because different frequencies, varying typologies of interactions between different persons, professions, organizations, stakeholders who have competing interest and goals (Liu 2009; Xie 2002).

Additionally, size, centrality, density and reachability have consistently been used to describe typical communication networks and linkages among participants (Tichy and Fombrun 1979; Mead 1999). This is referred to as the metrics of the communication network (Monge and Eisenberg 1987; Mead 1999). The size examines the number of persons linked in a communication network, whereas the centrality measures the volume of communication activities at a given communication node in the network (Liu 2009; Monge and Eisenberg 1987). Density refers to the ratio of actual to potential contacts within the network, whereas reachability focuses on the number of steps it takes to reach other persons in the network (Monge and Eisenberg 1987). The media for communication which refers to the tool or the technology that is used to transmit the conceived information or message are also deemed an important element in construction communication process (Dainty et al. 2006; Anumba and Evbuomwam 1999; Emmitt and Gorse 2003; Xie 2002). In the opinion of Xie (2002) and Emmitt and Gorse (2003), the construction industry adopts myriads of communication media (tools and technology), and these may take the form of postal letters, fax, telephone, face-to-face meetings or conversations, electronic mails, video conferences and in recent time virtual platforms such as BIM, augmented reality. These forms of media have been classified as oral, written and non-verbal communication (Shan 2003; Zhao 2003; Hunter 1993; Xie 2002). It is asserted that the richness of the media influences the effectiveness of the communication. However, media richness and lean properties vary from context, procurement model and the nature of interactions required (Manninen 2003; Otter and Emmitt 2007).

2.5 Dynamics of Construction Communication

Communication during across the life cycle phases of construction projects can have significant impact on performance, team effectiveness and overall success. Construction communication arises as the result of needed social and technical interactions and information exchanges between project actors (Xie et al. 2010; Emmitt and Gorse 2007; Otter and Emmitt 2007; Otter 2005; Thomas et al. 1998). The dynamics of communication and interactions among participants in a construction project delivery that underpinned key tasks and aspects of the delivery process are considered critical. According to Emmitt and Gorse (2007) and Hoezen et al. (2010), construction communication has taken varied forms from intra-supplier communication within the construction phase to demand–supply communication during the design phase. There has also been communication between and within single demand and supply side parties during the whole construction phase (Hoezen et al. 2010).

The demand side mostly has representatives such principals, users, investors, etc., whereas the supply side is made up of architects, engineer, contractors, subcontractors, advisors, project managers, etc. (Hoezen et al. 2010; Thomas et al. 1998, 1999). The need to adapt communication strategies, media, approaches, tools and technologies, skills, etc., to the nature of the context, tasks, complexities, networks and linkages, procurement models, project attributes has long been emphasized in several academic scholarships on communication performance (see Dainty et al. 2006; Kwofie et al. 2016; Kwofie 2015; Thomas et al. 1998; Xie 2002; Liu 2009; Xie et al. 2010). Hassall (2009) revealed that generally construction participants normally display similar patterns and profiles of communication on tasks even though the tasks had different communication requirements.

Hassan (1996) revealed five communication profile dynamics that are deemed significant in bringing about teamwork effectiveness and performance, and these were orientation, problem definition, criterion development, solution development and solution evaluation communication. Salazar (1997) operationalized orientation communication as the form that functions to make known a team's operating procedures and logistics and further determines the approach to resolving problems. Hassall (2009) on the other hand defined problem definition communication as one that aims to identify and define the nature and attributes of teams' task problems and possible causes. Criteria development communication however focuses on the values employed in evaluating decisions options or goals, whereas solution development communication concerns concrete, particular, specific proposal for action (Salazar

1997; Hassall 2009; Hirokawa and Salazar 1999). Solution evaluation communication indicates an evaluation of the positive and negative consequences of an adopted proposal (Hassall 2009; Hirokawa and Salazar 1999).

From Hassall (2009), it was noted that there are varying clusters of communication profile dynamics among project teams in construction project delivery. Thus, in some situations, project teams show the highest amount of communication activities in the solution development category then orientation category and equal volumes for problem definition and solution evaluation categories. Such a communication dynamics profile can be described as a team that utilizes significant time on communication concrete specific proposals for action that will solve a problem or shape how a task is solved (Hassall 2009). In other scenario, Hassall (2009) revealed that project teams tend to exhibit balanced orientation and problem definition profiles with a slightly less solution evaluation communication. This is an evidence a fairly active team in communication across the phases of the construction process.

From all these, poor communication is still a common feature of the construction industry and across the various life cycle phases. Consequently, despite all of the emphasis, effective communication and knowledge exchange between design and construction participants still seem to be a problem in practice. Against this, it can be argued that a more understanding of the communication profile dynamics that suits project types and contexts, procurement models, team dynamics must be rigorously pursued. Through this, a more bespoke profile would be revealed that can provide practical evidence on how to better integrate communication strategies and profiles into communication management practices and models to achieve improved performance outcomes.

2.6 The Context as a Key Factor in Construction Communication

Dainty et al. (2006) and Emmitt and Gorse (2007) contended that construction project environment presents unique context for communication that inherently requires both social and task-based interactions to deliver a set of tasks that realize some value and requires creating effective and seamless communication networks. The context of communication in construction environment can also be said to vary from one procurement model to the other. For example, Dainty et al. (2006) and Liu (2009) context for communication in traditional procurement significantly differs from that of non-traditional procurement models. In the opinion of Pietroforte (1997, across the six phases of construction project delivery process, there is an interplay of various participants which change across each phase and form a complex network and linkages in their communication and interactions. Salleh (2008) alluded that every context of communication will require varying degrees and different typologies of communication behaviours as well as experience different communication challenges.

Lailawati (2006) intimated that the judgement of appropriateness of communication task, tools or behaviour is contextually, situationally and culturally bound and as such an appropriate communication behaviour or strategies in one situation may not apply to another. However, with a slight adaptation and modification, a particular set of behaviours, practices, tools and technologies and approaches may seem appropriate to a given context (Salleh 2008; Wilson and Sabee 2003). Spitzberg and Cupach (2002) revealed that an individual's interpersonal skills along with his or her knowledge and motivation allow to trigger the requisite response to a given communication context. From this, the works of Emmitt and Gorse (2007) and Otter and Emmitt (2007, 2008) proved that the interactions and communication that take place at construction meetings and communications differ significantly from other working environment.

In recent times, it can be said that the construction environment for communication is changing and becoming varied (see Liu 2009; Emmitt and Gorse 2007; Otter and Emmitt 2007; Dainty et al. 2006). Shohet and Frydman (2003) revealed that a typical feature of construction environment in recent times is in the area of increasing integration, collaboration and reduction in time duration of projects by adopting concurrent activities. This drive in no doubt offer presents variabilities in communication requirements, frequencies, strategies networks and linkages that could be seen as unique to project environment. It is also a proven knowledge that construction projects are generally one-off and temporary in nature with participants mostly rely on project-specific and temporary networks of interactions to share knowledge, information and viewpoints when and where required in the project (Shohet and Frydman 2003; Chinowsky et al. 2008; Peng 1994). In this regard, Chinowsky et al. (2008) asserted that the nature of construction environment makes a need for interaction networks to support the efficient and free flow of knowledge, information, and ideas between project participants to achieve high performance through collective efforts an eminent necessity.

Furthermore, it is generally accepted that construction industry has witnessed continuous growth in the complexity of construction methods and techniques that are inherent in design and construction processes becoming progressively more sophisticated (Austin et al. 1997). This results in an interplay of a scenario requiring many specialist parties contribute to the technical aspects of the design and construction process thus creating a complex Web of inter-organizational relationships and critical information dependencies between project participants that is different from other industries such as manufacturing, production and services sectors (Pietroforte 1997; Sigband and Bell 1989). To this end, a good understanding of the complex construction project environment and its antecedent requirement for effective communication is a valuable asset towards engendering the needed improvement in interactions and information exchanges in project delivery.

2.7 Summary

This chapter presented an overview of the theoretical foundations of communication and the context of the construction project environment. An understanding of the dynamics of construction communication as well as the key elements in a communication chain is deemed very significant. This chapter posits that a need to engender effective communication performance must be led by the aligning of communication strategies and concepts to the unique attributes of the context and management model adopted the process and delivery. The next chapter focuses on non-traditional procurement and presents a sound case for the need to align communication to the notable attributes of such procurement model so as to enhance improvement.

References

Adler, R., Rosenfield, L., & Russell, P. (2004). Interplay: The process of interpersonal communication. *International Journal of Advertising, 29*(1), 85–110.

Anumba, C. J., & Evbuomwan, N. F. O. (1999). A taxonomy for communication facets in concurrent life-cycle design and construction. *Computer-Aided Civil and Infrastructure Engineering, 14*(1), 37–44.

Austin, S. A., Baldwin, A. N., & Newton, A. (1997). Manipulating the flow of design information to improve the programming of building design. *Construction Management and Economics, 12*, 445–455.

Brown, S. A. (2001). *Communication in the design process*. London: Spon press.

Cheng, T., & Teizer, J. (2013). Real-time resource location data collection and visualization technology for construction safety and activity monitoring applications. *Automation in Construction, 34*, 3–15.

Chinowsky, P., Diekmann, J., & Galotti, V. (2008). Social network model of construction. *ASCE, Journal Construction Engineering and Management, 134*(10), 804–812.

Cui, C., Liu, Y., Hope, A., & Wang, J. (2018). Review of studies on the public–private partnerships (PPP) for infrastructure projects. *International Journal of Project Management, 36*(5), 773–794.

Dainty, A., Moore, D., & Murray, M. (2006). *Communication in construction-theory and practice*. Milton Park, Abingdon, Oxon: Taylor & Francis 2 Park Square.

Du, D. (2004) *Information management*. Beijing: Tsinghua University press China.

Eisenberg, E. M., Farace, R. V., Monge, P. R., Bettinghaus, E. P., Kurchner-Hawkins, R., Miller, K., & Rothman, L. (1985). Communication linkages in interorganizational systems. In B. Derrin & M. Voigt (Eds.) *Progress in communication sciences* (Vol. 6, pp. 231–261). New York: Ablex.

El-Saboni, M., Aouad G., & Sabouni, A. (2009). Electronic communication systems effects on the success of construction projects in United Arab Emirates. *Advanced Engineering Informatics, 23*(1), 130–138.

Emmitt, S., & Gorse, C. A. (2003). *Construction communication*. Oxford: Blackwell Publishing.

Emmitt, S., & Gorse, C. A. (2007). *Communication in construction teams*. London: Taylor & Francis.

Gorse, C. A., & Emmitt, S. (2007). Communication behaviour during management and design team meetings: A comparison of group interaction. *Construction Management and Economics, 25*, 1197–1213.

Gorse, C. A., Emmitt, S., & Lowis, M. (1999). Problem solving and appropriate communication medium selection. In *ARCOM Association of researchers in construction management, 16th Annual Conference*, Liverpool John Moore University (pp. 511–518).

Goulden, N. R. (1992). Theory and vocabulary for communication assessments. *Communication Education, 41*(3), 258–269.
Guevara, J. M., & Boyer, L. T. (1981). Communication problems within construction. *Journal of Construction Engineering, American Society of Civil Engineers, 107*(4), 552–557.
Hassall, S. L. (2009). *The relationship between communication and team performance: Testing moderators and identifying communication profiles in established teams.* Ph.D. Thesis, Queensland University of Technology, Brisbane, Australia.
Hassan, T. M. (1996). *Simulating information flow to assist building design management.* Unpublished Ph. D. thesis, Department of Civil and building engineering, Loughborough University, Loughborough, UK.
Henderson, L. S. (2004). Encoding and decoding communication competencies in project management—An exploratory study. *International Journal of Project Management, 22,* 469–476.
Henderson, L. S. (2008). The impact of project managers' communication competencies: Validation and extension of a research model for virtuality, satisfaction, and productivity on project teams. *Project Management Journal, 39,* 48–59.
Hirokawa, R. Y., & Salazar, A. J. (1999). Task group communication and decision-making performance. In L. R. Frey (Eds.), *The handbook of group communication theory and research* (pp. 167–191). Thousand Oaks, CA: Sage.
Hoezen, M., Rutten, J. V., Voordijk, H., & Dewulf, G. (2010). Towards better customized serviceled contracts through the competitive dialogue procedure. *Construction Management and Economics, 28*(11), 1177 1186.
Howard, R., Kiviniemi, A., & Samuelson, O. (1998). Surveys of IT in the construction industry and experience of the IT barometer in Sacndinavia. *ITcon, 3,* 47–59.
Hunter, J. M. G. (1993). *Communications in construction.* Unpublished M.sc. Dissertation, Department of Civil and Building Engineering, Loughborough University, UK.
Jablin, F. M., Cude, R. L., House, A., Lee, J., & Roth, N. L. (1994). Communication competence in organizations: Conceptualization and comparison across multiple levels of analysis. In L. Thayer & G. Barnett (Eds.), *Organization-communication: Emerging perspective* (Vol. 4, pp. 114–140). Norwood, NJ: Ablex.
Jablin, F. M., & Sias, P. M. (2001). Communication competence. In F. M. Jablin & L. L. Putnam (Eds.), *The new handbook of organizational communication: advances in theory, research, and methods* (pp. 819–864). Thousand Oaks, CA: Sage.
Kreps, G. L. (1989). *Organisational communication: Theory and practice*, 2nd edn. Longman: New York.
Kwofie, E. (2015). *Contribution of unique features of mass housing projects to project team communication performance.* unpublished Ph.D. thesis, Kwame Nkrumah University of Science and Technology, Department of Building Technology, Kumasi.
Kwofie, T. E., Fugar, F., & Adinyira, E. (2016). Modelling the effect of housing design unit contract packaging on mass housing project team communication performance. *Journal of Construction in Developing Countries, 21*(1), 35–50.
Kwofie, T. E., Fugar, F. D. K., Adinyira, E., & Ahadzie, D. K. (2014). A conceptual framework for evaluating communication performance among mass housing project team. In O. Ejohwomu & O. Oshodi (Eds.), *Proceedings of 3rd International Conference on Infrastructure Development in Africa*, 17–19 March, 2014, Abeokuta, Nigeria (pp. 6–21).
Kwofie, T. E., Aigbavboa, C. O., & Thwala, W. D. (2019). Communication performance challenges in PPP projects: cases of Ghana and South Africa. *Built Environment Project and Asset Management, 9*(5), 628–641.
Lailawati, M. S. (2006). Characteristics of high/low context communication: The Malaysian Malay culture. *Journal of Intergroup Relations, 32*(4), 40–55.
Liu, Y. (2009). *Critical factors for managing project team communication at the construction stage.* Ph.D. Thesis. Submitted to the Polytechnic University of Hong Kong.

Manninen, T. (2003). Conceptual, communicative and pragmatic aspects of interaction forms-rich interaction model for collaborative virtual environments. In *The Proceedings of Computer Animation and Social Agents (CASA) Conference* (pp. 168–174). IEEE Computer Society Press.

Mead, S. T. (1999). *Communication effectiveness in intranet based construction projects*. Unpublished Ph. D. thesis, Department of Civil and building engineering, Loughorough, University, Loughorough, UK.

Monge, P. R., & Eisenberg, E. M., (1987). Emergent communication networks. In F. M. Jablin, L. L. Putnam, K. H. Roberts, & L. W. Porter (Eds.), *Handbook of organizational communication: An interdisciplinary perspective* (pp. 304–336).

Monge, P. R., Fulk, J., Kalman, M. E., Flanagin, A. J., Parnassa, C., & Rumsey, S. (1998). Production of collective action in alliance-based interorganizational communication and information systems. *Organization Science, 9,* 411–433.

Morreale, S. P., Spitzberg, B. H., & Barge, J. K. (2006). *Human communication: Motivation, knowledge, and skills* (2nd ed.). Belmont, CA: Thomson Learning.

Otter, A. F. H. J. (2005). *Design team communication using a project website.* PhD thesis, Bouwstenen 98, Technische Universiteit Eindhoven, Eindhoven.

Otter, A., & Emmitt, S. (2007). Exploring effectiveness of team communication Balancing synchronous and asynchronous communication in design teams. *Engineering, Construction and Architectural Management, 14*(5), 408–419.

Otter, A., & Emmitt, S. (2008). Design team communication and design task complexity: The preference for dialogues. *Architectural Engineering and Design Management, 4*(2), 121–129.

Peng, C. Z. (1994). Exploring communication in collaborative design: co-operative architectural modelling. *Design Studies, 15*(1), 19–44.

Perry, M., & Sanderson, D. (1998). Coordinating joint design work: The role of communication and artefacts. *Design Studies, 19,* 273–288.

Pietroforte, R. (1997). Communication and governance in the building process. *Construction Management and Economics, 15,* 71–82.

PMI. (2008). *A guide to the project management body of knowledge* (3rd ed.). Pennsylvania, USA: Project Management Institute.

Popple, G. W., & Towndrow, S. P. (1994). Communication for the construction industry. *BT Technology Journal, 13*(3), 45–50.

Richmond, V. P., & McCroskey, J. C. (1992). *Organisational communication for survival.* Englewood Cliffs, NJ: Prentice Hall.

Salazar, A. J. (1997). Communication effects on small group decision making: Homogeneity and task as moderators of the communication performance relationship. *Western Journal of Communication, 61,* 35–65.

Salleh, L. M. (2008). Communication competence: A Malaysian perspective. *Pacific and Asian Communication Association, 11*(3), 303–312.

Schermerhorn, J. R., Jr., Hunt, J. G., & Osborn, R. N. (1994). *Managing organisational behaviour.* New York, NY: Wiley.

Shan, X. H. (2003). *Communication studies, World and National.* Kunming: Yunnai University Press.

Shannon, C. E., & Weaver, W. (1949). *The mathematical theory of communication.* Chicago: University of Illinois Press.

Shohet, I. M., & Frydman, S. (2003). Communication patterns in construction at construction manager level. *Journal of Construction Engineering & Management, 129*(5), 570–577.

Sigband, N. B., & Bell, A. H. (1989). *Communication for management and business.* Glenview, Ill: Scott, Foresman and Co.

Sonnenwald, D. H. (1996). Communication roles that support collaboration during the design process. *Design Studies, 17*(3), 277–299.

Spitzberg, B. H., & Chagnon, G. (2009). Conceptualizing intercultural communication competence. In D. K. Deardorff (Ed.), *The SAGE handbook of intercultural competence* (pp. 2–52). Thousand Oaks, CA: Sage.

Spitzberg, B. H., & Cupach, W. R. (2002). Interpersonal skills. In M. L. Knapp & J. A. Daly (Eds.), *Handbook of interpersonal communication* (pp. 564–612). Thousand Oaks, CA: Sage.

Te'eni, D., Sagie, A., Zaidman, N., Schwartz, D., & Hamburger, Y. (2001, March). Organizational communication: A model and field study. *IEEE Transactions on Professional Communications,* 44(1).

Thomas, S. R. (1996). *An assessment tool for improving project team communication.* Ph.D. Thesis, University of Texas at Austin, USA.

Thomas, S. R., Tucker, R., & Kelly, W. (1998). Critical communications variables. *Journal of Construction Engineering Management, 129*(1), 58–66.

Thomas, S. R., Tucker, R., & Kelly, W. (1999). Compass: An assessment tool for improving project team communications. *Project Management Journal, 30*(4), 15–23.

Tichy, N. M., & Fombrun, C. (1979). Network analysis in organisational settings. *Human Relations, 32,* 923–965.

Weiner, B. (2006). *Social motivation, justice, and the moral emotions: An attributional approach.* Mahwah, NJ: Lawrence Erlbaum Associates.

Wikforss, Ö., & Lofgren, A. (2007). Rethinking communication in construction. *Journal of Information Technology in Construction, 12,* 337–346.

Wilson, S. R., & Sabee, C. M. (2003). Explicating communicative competence as a theoretical term. In J. O. Greene & B. R. Burleson (Eds.), *Handbook of communication and social interaction skills* (pp. 3–50). Mahwah, NJ: Lawrence Erlbaum.

Xie, X. (2002). *Communications in construction design.* Ph.D. thesis, Loughborough University, Loughborough.

Xie, C., Wu, D., Luo, J., & Hu, X. (2010). A case study of multi-team communications in construction design under supply chain partnering. *Supply Chain Management: An International Journal, 15*(5), 363–370.

Zhao, H. J. (2003). *Managing communication.* Beijing: Capital University of Economics and Business Press.

Part II
Understanding the Context of Non-Traditional Procurement Systems

Chapter 3
The Case of Non-traditional Procurement Systems and Communication

Abstract Non-traditional procurement systems are thought to offer significant benefits over conventional procurement arrangements in construction project delivery contracts; however, there is not much understanding of its dynamics and communication particularities. Non-traditional procurement models are continually evolving in response to changing market conditions. Communication issues are ranked high among supply chain relationships, complexity and client characteristics as the many factors that have a large impact on performance of non-traditional procurement models. Given that non-traditional procurement models are gaining increasing popularity in the global construction industry, it is thus paramount for deeper theoretical and conceptual insights into their attributes, dynamics and context of communication and information sharing. The focus of this chapter is on the key attributes of non-traditional procurement typologies, their context for effective communication and barriers experienced. These insights can form solid platform on which the benefits of non-traditional procurement systems for construction project delivery can be optimized. With the significance of communication and procurement attributes to project performance, the changing business environment characterized by tense competitiveness and wide global links places a demand on construction industry practitioners to establish effective and efficient integrated and inter-organization collaborative management systems through non-traditional procurement models. This is a sure way to improving performance and enhancing competitiveness by responding to the changing environment in construction project delivery.

Keywords Non-traditional procurement · Conventional procurement · Information sharing · Procurement attributes

3.1 Introduction

The need to optimize the benefits of non-traditional procurement typologies has overly been suggested and proposed as an area needing more research. Aligning communication concepts, management and strategies to the key attributes of non-traditional procurement is deemed as a panacea for enhancing the predictive accuracy of its benefits. This chapter presents an understanding of non-traditional procurement,

T. E. Kwofie et al., *Effective Construction Project Delivery*,
https://doi.org/10.1007/978-3-030-49374-5_3

its unique attributes, various types and the nature of communication inherent in such procurement models. The essence of this drive is underpinned by the theoretical foundation that effective communication has overly been ranked among the key factors that impact on the performance of non-traditional procurement.

3.2 Understanding Non-traditional Procurement

Non-traditional procurement models such as collaborative working, relational contracting, integrated project delivery, partnership and project alliance have gained credence and increasing acceptance and popularity as a sure procurement model thought to offer significant superior benefits over traditional contracts (Song et al. 2009; Bresnen 2010; Chan et al. 2010; Kent and Becerik-Gerber 2010; Hartmann and Bresnen 2011; Asmar et al. 2013; Xue et al. 2010; Manley and Chen 2016). According to Davis and Walker (2009), Rahman and Kumaraswamy (2012) and Miller et al. (2009) non-traditional procurement have a unique attribute in its ability to be continually evolving in response to changing market conditions. Xue et al. (2010) alluded that the changing business environment characterized by tense competitiveness and wide global links requires construction process and organizations to establish effective and efficient inter-organization integrated and collaborative management systems towards improving performance and success. Non-traditional procurement is deemed as an emerging working model for improving performance and enhancing competitiveness by responding to the changing environment in construction (Xue et al. 2010; Davis and Walker 2009; Kelly 2011; Bresnen 2010; Chan et al. 2010; Kent and Becerik-Gerber 2010; Hartmann and Bresnen 2011; Ling et al. 2015; Asmar et al. 2013; Manley and Chen 2016).

In the construction industry context, large volume of research scholarships and literature have given rise to a large number of definitions of the term procurement method, procurement model, procurement system or procurement route (see Love et al. 2010; Austin et al. 2001; Miller et al. 2009; Naoum 2011; Ali et al. 2011; Furneaux et al. 2008; Eriksson 2007; Davis et al. 2008; Eriksson 2017; Forgues and Koskela 2009; Hughes et al. 2012; Ruparathna and Hewage 2013; Shafik and Martin 2006; Morledge and Smith 2013). Two of the many definitions of procurement system or model that encapsulate the construction industry and project delivery practices are as presented by Love et al. (1998, p. 221) as:

> an organizational system that assigns specific responsibilities and authorities to people and organizations, and defines the relationships of the various elements in the delivery of a construction project. Love et al. (1998, p. 221) and

Naoum (2011, p. 2) as:

> a mechanism for linking and coordinating members of the construction team throughout the building process in a unique systematic structure, both functionally via roles, authority and power, and contractually via responsibilities and risks Functionally with the aim to deliver a project that meets its objectives and fulfill the client criteria and expectations.

The two broad classification of procurement systems has been traditional and non-traditional procurement (see Manley and Chen 2016; Hughes et al. 2012; Davis et al. 2008; Eriksson 2017; Song et al. 2009; Love et al. 2008; Bresnen 2010; Chan et al. 2010; Kent and Becerik-Gerber 2010). The term traditional procurement system is also referred to as the conventional procurement system which entails the completion of three main successive stages being design, tender and construction (build) (Turner 1997; Ashworth and Hogg 2007). In traditional procurement, a group of persons must be engaged to complete the design of the projects, prepare necessary building documents, select a group of contractors and subcontractors and then finally start the on-site construction activities in a linear or sequential manner (Hughes et al. 2012; Davis et al. 2008; Love et al. 2008; Naoum 2011; Ali et al. 2011; Furneaux et al. 2008; Eriksson 2007). A key attribute of the traditional procurement system or models is that each stage must be completed before the next stage can start. For example, the design and documentation of the project must be completed before conducting tender to select the contract which must also be completed before the start of on-site construction of the project. Drexler and Larson (2000) and Xue et al. (2010) revealed that traditional or conventional procurement systems are adversarial in nature.

Non-traditional procurement systems on the other hand have emerged as a response to construction industry initiatives and changes in client requirements providing a unified interface between the construction project and the employer's organization and ameliorating the adversarial relationship associated with conventional procurement models as well as enhancing collaboration (Xue et al. 2010; Hughes et al. 2012; Davis et al. 2008; Love et al. 2008; Bresnen 2010; Chan et al. 2010). Project alliancing, relational contracting, joint ventures, a common feature of non-traditional procurements are its integrated and collaborative nature, and this feature is recognized as one of the most significant criteria for success for operating construction (Xue et al. 2010; Bell 2003; Shen and Wu 2005; Wong and Hui 2006). According to Xue et al. (2010) which is shared by other researchers, non-traditional procurement systems cover a spectrum of working model through ways where two or more organizations or work teams work together through informal networks, alliances, partnering, joint ventures and full integration. The various forms of non-traditional procurements systems are teamwork, partnership, project alliance, joint venture, strategic alliance, coalition, etc. (Shen and Wu 2005; Wong and Hui 2006; Hauck et al. 2004; Pryke 2004; London and Kenley 2001; Anumba et al. 2003; Sillars and Kangari 2004; Beach et al. 2005; Fong and Lung 2007; Manley and Chen 2016).

A detailed account on these non-traditional procurement typologies is contained in chapter five. Additionally, the inherent suitability and factors for choice of these typologies of procurement models are beyond the scope of this book given that numerous research has adequately dealt with this subject matter (see Miller et al. 2009; Naoum 2011; Naoum and Egbu 2015; Hughes et al. 2012; Davis et al. 2008; Love et al. 2008; Ali et al. 2011; Furneaux et al. 2008; Eriksson 2007; Xue et al. 2010; Ruparathna and Hewage 2013).

3.3 Key Attributes and Nature of Communication in Non-traditional Procurement

There is a widespread recognition of urgent need to significantly improve on communication performance in non-traditional procurement due to the centrality of the role of communication in optimizing its inherent benefits (Kwofie 2015; Xue et al. 2010; Manley and Chen 2016). Open and frequent communication towards a more integrated and collaborative working is identified as an important component of a positive work norms and culture and an enabling factor for optimizing performance in non-traditional procurement models (Cigularov et al. 2010; Kwofie et al. 2019a). It is contended that communication in non-traditional procurement systems is more likely to be effective when it is multi-directional in information sharing and social and task-based interactions (Kwofie et al. 2019b; Manley and Chen 2016; Chen and Manley 2014). The non-adversarial, integrated and collaborative nature of non-traditional procurement highlight the critical role played by communication in establishing consensual group norms especially in social interactions (Rimal and Real 2003; Tai et al. 2009; Ling et al. 2015; Chan et al. 2008). This is because both tasks-based and social interactions which are norm-based can occur only through communication as norms cannot exist in the absence of communication among members of a group or work teams (Rimal and Real 2003).

It can be adduced that communications and interactions in various non-traditional procurement systems in construction projects are a very complex issue and sometimes are very challenging to comprehensively understand the whole picture. Kwofie et al. (2019a) and Manley and Chen (2016) alluded that non-traditional procurements form complex communication networks due to the hybrid formal and informal governance structures that are adopted to manage transactions.

Generally, a common feature of non-traditional procurement systems is the tendency of overlapping the design and construction suggesting that often, the project cannot be designed in a logical order (Xie et al. 2010; Ling et al. 2015; Tai et al. 2009). Accordingly, communication, including integration of specialized knowledge and negotiation of differences between multi-team members, has emerged as a fundamental component of the design and construction processes (Sonnenwald 1996; Ling et al. 2015). Additionally, there is also the incidence of many actors and participants forming the project teams in the construction supply chain from widely distributed geographic areas that must interact and share project information effectively to enhance integrated task function and seamless collaboration (Xie et al. 2010; Xue et al. 2010).

Tai alluded that, in typical non-traditional procurement models, there is the incidence of many and varied participants with different communication channels, and it is near impossible to effectively share information and interact uniformly by a single communication model or strategies. There is also an ever-changing participant who adopts varying forms of communications that are required to complete their tasks (Tai et al. 2009; Ling et al. 2015). Predominantly, there are also various organizations who exhibit different organizational structures, and these communicate through specific individuals resulting in more variables to their construction project communications (Manley and Chen 2016; Tai et al. 2009; Chen et al. 2012).

In Ling et al. (2015) and Tai et al. (2009) face-to-face meeting, letters, telephone, fax and email have been normally used as dominant communication media in most partnerships and project alliancing projects in China and Hong Kong. Following this, Xie et al. (2010) alluded that the general global construction industry has given extensive increased attention to the technical aspects of multi-team communication who are mostly colocated or distributed. In this regard, many new technologies have been adopted in order to enhance interactions and achieve good coordination and communication between distributed and colocated project teams (Perry and Sanderson 1998; Wikforss and Lofgren 2007). These technologies are aimed at bridging the virtual gap among project teams. From this, it can be affirmed that the attributes and nature of communication in non-traditional procurement systems range from oral, textual, digital and non-verbal in both social and task-based interactions.

3.4 Challenges and Barriers in Non-traditional Procurement

From extant literature, supply chain relationships, site conditions, regulation, risk, information asymmetry, complexity and client characteristics are suggested as key factors that significantly impact on performance of non-traditional procurement (Lai et al. 2008; Rahman and Kumaraswamy 2008; Manley and Chen 2016; Osipova and Eriksson 2011). These factors among others are also thought impact of project outcomes and team dynamics through the adopted governance mechanisms, shape the scope for goal achievement and determine innovation potential (Chen and Manley 2014; Müller and Martinsuo 2015).

Construction project delivery process evolves through discrete phases meaning each phase of which can be viewed as a discrete transaction during which key participants and stakeholders rotate in and out of the project (Chen and Manley 2014; Müller and Martinsuo 2015; Chen et al. 2018). This attribute is said to present notable challenges and barriers to risk management, communication management and information sharing irrespective of the procurement typologies. However, Chen et al. (2018) and Xue et al. (2010) suggested that this attribute presents more challenges to governance, information sharing, team dynamics, network complexities and innovation in non-traditional procurement such as relational contracting, integrated, collaborative working and partnerships which are due to the unique human and business environment factors.

It has also been overly proven that there is high incidence of unpredictability on project outcomes on non-traditional procurement which is often due to the characteristics of clients (Joslin and Müller 2016; Müller and Martinsuo 2015; Chen et al. 2018).

Blanc-Brude (2008) and Luo (2007) intimated that existence of informational asymmetries among the government (principal) and the private firm (agent) in partnership working model remains an immediate challenge. Against this, Ceric (2011) and De-Palma et al. (2009) have advised that information asymmetries must be factored in during the design of any partnership arrangements. Often in partnership contract working models, the agent is better informed than the principal on related ongoing activities and the actions of the private party that impact on these activities resulting in severe information asymmetries (Blanc-Brude 2008).

Yang and Yang (2010) and Biong (2013) also contended that, in most non-traditional procurement models especially in partnerships, relational contracting and joint ventures, Moral hazard and adverse selection problems are obvious challenges which do not exist in isolation. It is argued that these are normally inherent from "the existence of synergies between stages of the venture, accounting for why different tasks are bundled in a distinct activity and delegated to a sole responsible agent firm" (Biong 2013, p. 6). In this regard, Blanc-Brude (2008) Luo (2007) also revealed that the efforts of sole agent firm exert numerous influences and circumstances they encounter at the operation stage. It is alluded that non-traditional procurement is usually based on relationship building, but Müller and Martinsuo (2015) relational norms in the buyer–supplier relationships are moderated by the strictness of project governance, especially the level of flexibility left to the project manager. However, lower levels of managerial flexibility are detrimental to project success in cases of weak relational norms and supportive of project success in cases of high relational norms are positively associated with project success (Müller and Martinsuo 2015; Chen et al. 2018; Chen and Manley 2014). In the revelation provided by Müller and Martinsuo (2015), clear organizational structures and methodologies are supportive of project success in cases of good relational norms. However, it is said that non-traditional procurements have relatively diffused relational norms and flexible contractual relationship which induce some amount of influences and challenging circumstances on relational and governance performance.

There are also incidences of social and intangible features of relational norm that characterize the relationship in many typologies of relationships (i.e. supplier and a buyer, client and sub-contractor, project manager and contractor) in most non-traditional procurement models (Xue et al. 2010). Wu et al. (2016) established that non-traditional procurement such as joint ventures and integrated project delivery exhibits flexible accountability systems and modalities that are by far in sharp contrast to traditional procurement, and this must be carefully managed if performance and teamwork are to be effective. According to Müller and Turner (2005) and Napier et al. (2009), organic governance is more suited to non-traditional procurement, and they normally provide high flexibility in managerial actions which does not influence variance in relational norms. To this end, it can be contended relational norms are precursor for effective teamwork and task function in relational contracting and collaborative working. Thus, high relational norms can significantly impact the project success (Müller and Turner 2005; Napier et al. 2009).

3.5 Summary

Owing to urgent and emerging performance necessity for optimizing the benefits of non-traditional procurement systems, gaining and in-depth understanding of the key attributes of non-traditional procurement typologies and their inherent challenges and barriers are significant. Clients, project team participants and other stakeholders who form the nucleus of the communication networks and channels will continually be expected to ensure effective information sharing and communication that can engender the needed performance for optimized benefits. This is because communication has been overly suggested as one of the critical factors in achieving this. This chapter has adequately espoused the attributes and notable challenges of non-traditional procurement systems. In so doing, a clear picture that distinctly differentiates it from traditional procurement systems has been created.

References

Ali, Z. A., Zakaria, N., & Che-Ani, A. I. (2011). The effect of procurement system towards the performance of refurbishment works. *America*, pp. 70–75.

Anumba, C. J., Ren, Z., Ugwu, O. O., Thorpe, A., & Newnham, L. (2003). Negotiation within a multi-agent system for the collaborative design of light industrial buildings. *Advances in Engineering Software, 34*(7), 389–401.

Ashworth, A., & Hogg, K. (2007). *Willis's practice and procedure for quantity surveyor*. Oxford: Blackwell Publishing Ltd.

Asmar, M., Hanna, A., & Loh, W. (2013). Quantifying performance for the integrated project delivery system as compared to established delivery systems. *Journal of Construction Engineering and Management, 139*(11), 4013012-1–4013012-14.

Austin, S. A., Baldwin, A. N., Hammond, J., Murray, M., Root, D., Thomson, D., & Thorpe, A. (2001). *Design chains: A handbook for integrated collaborative Design*. Thomas Telford Publishing, London.

Beach, R., Webster, M., & Campbell, K. M. (2005). An evaluation of partnership development in the construction industry. *International Journal of Project Management, 23*(8), 611–621.

Bell, S. (2003). *Economic governance and institutional dynamics*. Melbourne: Oxford University Press.

Biong, H. (2013). Choice of subcontractor in markets with asymmetric information: Reputation and price effects. *Journal of Business & Industrial Marketing, 28*(1), 60–71.

Blanc-Brude, F. (2008). *Public-private risk transfer—Public-private partnerships, long-term contracts and price discovery*. London: King's College.

Bresnen, M. (2010). Keeping it real? Constituting partnering through boundary objects. *Construction Management & Economics, 28*(6), 615–628.

Ceric, A. (2011). Minimizing communication risk in construction projects: A Delphi study of the key role of projects managers. Working Paper Proceedings, Engineering Projects Organizations Conference, Estes Park, Colorado, 9–11 August.

Chan, A., Chan, D., Fan, L., Lam, P., & Yeung, J. (2008). Achieving partnering success through an incentive agreement: Lessons learnt from an underground railway extension project in Hong Kong. *Journal of Management in Engineering, 24*(3), 128–137.

Chan, A. P. C., Chan, D. W., & Yeung, J. F. (2010). *Relational contracting for construction excellence: Principles, practices and case studies*. Abingdon: Spon Press.

Chen, W. T., Chen, T. T., Sheng, L. C., & Liu, S.-S. (2012). Analyzing relationships among success variables of construction partnering using structural equation modelling: A case study of Taiwan's construction industry. *Journal of Civil Engineering and Management, 18*(6), 783–794.

Chen, L., & Manley, K. (2014). Validation of an instrument to measure governance and performance on collaborative infrastructure projects. *Journal of Construction and Engineering Management, 140*(5), 04014006. https://doi.org/10.1061/(ASCE)CO.1943-7862.0000834.

Chen, L., Manley, K., Lewis, J., Helfer, F., & Widen, K. (2018). Procurement and governance choices for collaborative infrastructure projects. *Journal of Construction and Engineering Management, 144*(8), 04018071.

Cigularov, K. P., Chen, P. Y., & Rosecrance, J. (2010). The effects of error management climate and safety communication on safety. *A Multi-Level Study. Accident Analysis and Prevention, 42*(5), 1498–1506.

Davis, P. R., Love, P. E. D., & Baccarini, D. (2008). Building procurement selection review. Report No. 1 (Project no. 2006.034-C report no. 1) For the CRC in Construction Innovation, Australia.

Davis, P. R., & Walker, D. H. T. (2009). Building capability in construction projects: A relationship-based approach. *Engineering, Construction and Architectural Management, 16*(5), 475–489.

De-Palma, A., Leruth, L., & Pruneir, G. (2009). Towards a principal-agent based typology of risks in public-private partnerships, IMF. Available at: www.imf.org/en/Publications/WP/Issues/2016/12/31/Towards-a-Principal-Agent-Based-Typology-of-Risks-in-Public-Private-Partnerships-23190. Accessed May 11, 2019.

Drexler, J. A., & Larson, E. W. (2000). Partneering: Why project owner-contractor relationships change. *Journal of Construction Engineering and Management, ASCE, 126*(4), 293–297.

Eriksson, P. E. (2007). The influence of partnering and procurement on subcontractor involvement and innovation. *Facilities, 25*(5/6), 203–214.

Eriksson, P. E. (2017). Procurement strategies for enhancing exploration and exploitation in construction projects. *Journal of Financial Management of Property and Construction, 22*(2), 211–230.

Fong, P. S. W., & Lung, B. W. C. (2007). Inter-organizational teamwork in the construction industry. *Journal of Construction and Engineering Management, 133*(2), 157–168.

Forgues, D., & Koskela, L. (2009). The influence of a collaborative procurement approach using integrated design in construction on project team performance. *International Journal of Managing Projects in Business, 2*(3), 370–385.

Furneaux, C. W., Brown, K., & Allan, D. (2008). Public values embedded in Australian public works procurement. *Public Money and Management, 28*(3), 167–172.

Hartmann, A., & Bresnen, M. (2011). The emergence of partnering in construction practice: An activity theory perspective. *Engineering Project Organization Journal, 1*(1), 41–52.

Hauck, A. J., Walker, D. H. T., Hampson, K. D., & Peters, R. J. (2004). Project alliancing at national museum of Australia—Collaborative process. *Journal of Construction and Engineering Management, 130*(1), 143–152.

Hughes, D., Williams, T., & Ren, Z. (2012). Is incentivisation significant in ensuring successful partnered projects null. *Engineering, Construction and Architectural Management, 19*(3), 306–319.

Joslin, R., & Müller, R. (2016). The relationship between project governance and project success. *International Journal of Project Management, 34*(1), 613–626.

Kelly, J. (2011). *Cracking the VFM code: How to identify & deliver genuine value for money in collaborative contracting*. Sydney: Big Fig Publishing Limited, under the imprint of Intelligentsia Press.

Kent, D. C., & Becerik-Gerber, B. (2010). Understanding construction industry experience and attitudes toward integrated project delivery. *Journal of Construction Engineering and Management, 136*(8), 815–825.

Kwofie, T. E. (2015). *Contribution of unique features of mass housing projects to project team communication performance*. Unpublished PhD thesis, Kwame Nkrumah University of Science and Technology, Kumasi.

Kwofie, T. E., Aigbavboa, C. O., & Thwala, W. D. (2019a) Communication performance challenges in PPP projects: cases of Ghana and South Africa. *Built Environment Project and Asset Management, 9*(5), 628–641.

Kwofie, T. E., Aigbavboa, C. O., & Machethe, S. O. (2019b) Nature of communication performance in non-traditional procurements in South Africa. *Engineering, Construction and Architectural Management, 26*(10), 2264–2288.

Lai, Y., Wang, W., & Wang, H. (2008). AHP-and simulation-based budget determination procedure for public building construction projects. *Automation in Construction, 17*(5), 623–632.

Ling, F., Peng Chong, T., Yan, N., Albert, T., & Asanga, G. (2015). Effect of adoption of relational contracting practices on relationship quality in public projects in Singapore null. *Engineering, Construction and Architectural Management, 22*(2), 169–189.

London, K. A., & Kenley, R. (2001). An industrial organization economic supply chain approach for the construction industry: A review. *Construction Management and Economics, 19*(8), 777–788.

Love, P., Davis, P., Baccarini, D., Wilson, G., & Lopez, R. (2008). Procurement selection in the public sector: Atatle of two state, pp. 1–11.

Love, P. E. D., Mistry, D., & Davis, P. R. (2010). Price competitive alliance projects: Identification of success factors for public clients. *Journal of Construction Engineering and Management, 136*(9), 947–956.

Love, P. E. D., Skitmore, M., & Earl, G. (1998). Selecting a suitable procurement method for a building project. *Construction Management and Economics, 16*(2), 221–233.

Luo, Y. (2007). The independent and interactive roles of procedural, distributive, and interactional justice in strategic alliances. *Academy of Management Journal, 50*(3), 644–664.

Manley, K., & Chen, L. (2016). The impact of client characteristics on the time and cost performance of collaborative infrastructure projects. *Engineering, Construction and Architectural Management, 23*(4), 511–532.

Miller, G., Furneaux, C., Davis, P., Love, P., & O'Donnell, A. (2009). *Built environment procurement practice: impediments to innovation and opportunities for changes*. Curtin University of Technology.

Morledge, R., & Smith, A. (2013). *Building procurement* (2nd ed.). Oxford: Wiley-Blackwell.

Müller, R., & Martinsuo, M. (2015). The impact of relational norms on information technology project success and its moderation through project governance. *International Journal of Managing Projects in Business, 8*(1), 154–176.

Müller, R., & Turner, J. R. (2005). The impact of principal-agent relationship and contract type on communication between project owner and manager. *International Journal of Project Management, 23*(5), 398–403.

Naoum, S. (2011). *People and organizational management in construction*. London: Thomas Telford Publishers.

Naoum, S., & Egbu, C. (2015). Critical review of procurement method research in construction journals. *Procedia Economics and Finance, 21*, 6–13.

Napier, N. P., Kell, M., & Tan, F. B. (2009). IT project managers' construction of successful project management practice; A repertory grid investigation. *Information System Journal, 19*(3), 255–282.

Osipova, E., & Eriksson, P. E. (2011). How procurement options influence risk management in construction projects. *Construction Management & Economics, 29*(11), 1149–1158.

Perry, M., & Sanderson, D. (1998). Coordinating joint design work: The role of communication and artefacts. *Design Studies, 19*(3), 273–288.

Pryke, S. D. (2004). Analysing construction project coalitions: Exploring the application of social network analysis. *Construction Management and Economics, 22*(8), 787–797.

Rahman, M. M., & Kumaraswamy, M. M. (2008). Relational contracting and teambuilding: Assessing potential contractual and non-contractual incentives. *Journal of Management in Engineering, 24*(1), 48–63.

Rahman, M. M., & Kumaraswamy, M. M. (2012). Multicountry perspectives of relational contracting and integrated project teams. *Journal of Construction Engineering and Management, 138*(4), 469–480.

Rimal, N. R., & Real, K. (2003). Understanding the influence of perceived norms on behaviours. *Communication Theory, 13*(2), 184–203.

Ruparathna, R., & Hewage, K. (2013). Review of contemporary construction procurement practices. *Journal of Management Engineering*. https://doi.org/10.1061/(ASCE)ME.1943-5479.0000279, 04014038.

Shafik, M., & Martin, P. (2006). The impact of procurement methods on the Scottish house building industry. In *22nd Annual ARCOM Conference*, 4–6 September 2006, Birmingham, UK, Association of Researchers in Construction Management (pp. 81–90).

Shen, L. Y., & Wu, Y. Z. (2005). Risk concession model for build/operate/transfer contract projects. *Journal of Construction and Engineering Management, 131*(2), 211–220.

Sillars, D. N., & Kangari, R. (2004). Predicting organizational success within a project-based joint venture alliance. *Journal of Construction and Engineering Management, 130*(4), 500–508.

Song, L., Mohamed, Y., & AbouRizk, S. M. (2009). Early contractor involvement in design and its impact on construction schedule performance. *Journal of Management in Engineering, 25*(1), 12–20.

Sonnenwald, D. H. (1996). Communication roles that support collaboration during the design process. *Design Studies, 17*(3), 277–299.

Tai, S., Wang, Y., & Anumba, C. J. (2009). A survey on communications in large-scale construction projects in China. *Engineering, Construction and Architectural Management, 16*(2), 136–149.

Turner, A. (1997). *Building procurement*. London: Palgrave Macmillan Ltd.

Wikforss, Ö., & Lofgren, A. (2007). Rethinking Communication in Construction. *Journal of Information Technology in Construction, 12*, 337–346.

Wong, J. T. Y., & Hui, E. C. M. (2006). Construction project risks: Further considerations for constructors' pricing in Hong Kong. *Construction Management and Economics, 24*(4), 425–438.

Wu, J., Liu, J. H., Xiaohua, J., & Sing, M. C. P. (2016). Government accountability within infrastructure public-private partnerships. *International Journal of Project Management, 34*(8), 1471–1478.

Xie, C., Wu, D., Luo, J., & Hu, X. (2010). A case study of multi-team communications in construction design under supply chain partnering. *Supply Chain Management: An International Journal, 15*(5), 363–370.

Xue, X., Shen, Q., & Ren, Z. (2010). Critical review of collaborative working in construction projects: Business environment and human behaviors. *Journal of Management in Engineering, 26*(4), 196–208.

Yang, J. B., & Yang, C. C. (2010). Evaluating schedule delay causes for private participating public construction works under the build-operate-transfer model. *International Journal of Project Management, 28*(1), 569–579.

Chapter 4
Procurement Systems: Underscoring Their Particularities

Abstract Various forms of procurement systems and arrangements in project delivery have increasingly become popular around. In spite of arguments in favour of other over others due to their potential to provide improved project performance, project outcomes and performance continue to be unpredictable regardless of the rational for the choice of the procurement type. Several criteria have formed the basis of the choice of particular procurement typologies. However, there is lack of understanding of the inherent particularities of these procurement typologies and their impact on performance, management, team effectiveness, governance and communication. An in-depth understanding of these attributes will form a valuable information and springboard for clients in performing their single most important task of choosing the most suitable and appropriate procurement system for their project delivery. Still clients and project teams rely on knowledge of cost and time implication for choice of procurement typologies without an understanding of their performance due to their attributes. This needs to change by client focusing more on the attributes of procurement typologies and its implications for performance, management, governance and communication as this drive will undoubtedly impact on the uncertainties already associated with them. This chapter delineates the inherent attribute of the two main procurement groups and provides an insight into their inherent communication implication.

Keywords Project systems · Project delivery · Procurement attributes · Project performance · Project outcomes

4.1 Introduction

Choosing the most appropriate and context suitable procurement model for a project delivery remains one of the most significant responsibilities of the client and project teams. In this regard, a knowledge of the attributes and performance characteristics of the array of procurement models and systems is deemed critical to the choice and decisions of the project teams. Various procurement typologies have been espoused and used in project delivery in the construction industry amidst heightened uncertainties of its performance and benefits. Lack of understanding of the inherent attributes

T. E. Kwofie et al., *Effective Construction Project Delivery*,
https://doi.org/10.1007/978-3-030-49374-5_4

of these procurement typologies makes further increase the uncertainties in their adoption as these attributes have implications for management, team effectiveness, performance and communication. This chapter focuses on the concept of typologies of procurement systems normally adopted in the construction industry. An attempt is made to delineate the attributes of the two main groups of procurement in construction project delivery as well as the implications of these attributes on communication performance.

4.2 Typologies of Procurement Systems in Construction

The type of procurement model or system adopted in the delivery of construction projects is said to have significant influence on both efficiency-related and innovation-related problems (Assaf and Al-Hejji 2006; Shehu et al. 2014; Tawiah and Russell 2008; Bosch-Sijtsema and Postma 2009). According to Hughes et al. (2006), the selection of an appropriate procurement method will assist in obtaining value for money and should make effective use of resources. However, this is possible with the in-depth understanding of the typologies of procurement systems and their inherent particularities. The theoretical foundation of this assertion is the fact that each procurement system is unique and this uniqueness has significant implication for practice and theory in order to optimize its benefits and achieve better project performance.

Indeed the term "construction procurement method" has been variedly defined in plethora of existing literature. According to Chan et al. (2007), construction procurement method is defined as the system that represents the organizational structure adopted by clients for the implementation of project processes and eventual operation of the project. Molenaar et al. (2009) on the other hand defined construction procurement method as a comprehensive process through which designers, constructors and various project participants and consultants provide services for design, construction and management to deliver a project to the client. Palaneeswaran et al. (2003) also referred to construction procurement as the process of acquiring or obtaining material, property or services, begins with the determination of a need for a property or service and ends with the completion and closeout of a contract. Love et al. (1998: p. 222) on the other hand defined it as an organizational system that assigns specific responsibilities and authorities to people and organizations and defines the various elements in the construction of a project.

From these definitions, it can be clearly deduced that there are wide range of processes involved in a procurement strategy or method. Many also suggest that these processes are indeed often interrelated and sequential in nature requiring effectiveness and efficiency which impact considerably on the success or performance of the projects (Latham 1994; Palaneeswaran et al. 2003; Kumaraswamy and Dissanayaka 2001; Luu et al. 2005). Exploring various procurement methods in the construction industry from the turn of the 1980s was necessitated by the need for urgent changes in the traditional patterns of relationships between the professions in project delivery towards a more collaborative and integrated working approach (Palaneeswaran et al. 2003; Kumaraswamy and Dissanayaka 2001; Emmitt and Gorse 2007). This was because the changing environment required a different philosophy as cooperation, good channels of communication and good governance became paramount in ensuring the successful completion of a project (Latham 1994; Xue et al. 2010; Chan et al. 2007; Luu et al. 2005; Love et al. 2008).

Plethora of existing literature has used various terms such as procurement methods, project approach, project delivery systems or procurement delivery methods to refer to procurement system in the construction industry (Turner 1997; Ashworth and Hogg 2007; Ranganathan and Lerpittayapoom 2002, 2006; Naoum and Egbu 2015; Eriksson 2017; Morledge and Smith 2013; Love et al. 2008). Procurement systems in construction project delivery have consistently been classified into two main types, namely traditional and non-traditional procurement (see Turner 1997; Xue et al. 2010; Ling et al. 2015; Manley and Chen 2016; Naoum and Egbu 2015; Eriksson 2017; Morledge and Smith 2013; Love et al. 2008; Ashworth and Hogg 2007). The traditional procurement system is also referred to as conventional procurement system (Seeley 1997; Turner 1997).

The traditional procurement systems entail the strict compliance to the following successive stages entailing design, tender and build or construction which are linear and sequential (Naoum and Egbu 2015; Eriksson 2017; Ashworth and Hogg 2007). Conventionally, in this type of procurement system, the preceding stage will have to be completed before the start of the next stage since the product of later stage is needed to initiate proceeding stage. The non-traditional procurement systems emerged as a response to the emerging dynamics of the global market conditions, the complexities of the construction environment, amplified scope and intricacies of the building schemes and process, economic encounters, social and political thoughtfulness so as to improve performance and enhancing competitiveness (Manley and Chen 2016; Song et al. 2009; Bresnen 2010; Chan et al. 2010; Morledge and Smith 2013). Non-traditional procurement systems have been deemed as a responsive strategy to overcoming the adversarial nature of construction by stimulating collaboration between parties so as to attain mutual goals (Li et al. 2000).

4.3 Understanding the Inherent Attributes of Construction Procurement Typologies

Most of existing procurement systems often employed in the construction industry share unique attributes and features which have implications for management, information sharing and relationship management (Xue et al. 2010; Chan et al. 2010; Morledge and Smith 2013; Li et al. 2000). Consistently, business environment and human factors have been used to delineate the unique attributes of many management styles, concepts, procurement systems especially in construction (Chen 2003; Lau et al. 2004; Green et al. 2005; Cicmil and Marshall 2005). The business environment of construction projects especially in traditional procurement systems is usually characterized by adversarial relationships, fragmented operation processes, a lack of genuine cooperation over time and complexity (Cicmil and Marshall 2005; Xue et al. 2010; Beach et al. 2005). The business environment in non-traditional procurement systems is said to be integrative, collaborative and cooperative relationship, flexible and exhibits relational contracting norms that hold potential in enhancing construction project outcomes (Kumaraswamy et al. 2005; Xue et al. 2010; Ning 2014).

Additionally, Xue et al. (2010) and Cicmil and Marshall (2005) alluded that traditional procurement systems are mostly characterized by rigid contracting frameworks for the sake of accountability and high risk sharing and allocation. In non-traditional procurement systems however, there is a more sense towards flexible contracting that ameliorates the often adversarial environment often associated with construction project delivery. Xue et al. (2010) further alluded that collaborative problem-solving, commitment and agreement, common value, long-term relationship, clear definition of responsibilities are the core business operations peculiar to non-traditional procurement systems. In the considered opinion of Ling et al. (2015), traditional procurement systems often exhibit rigid organizational and team culture, variant attitudes, rigid code of conduct and high competing goals and expectation which guide behaviour, whereas the opposite is the case in non-traditional. In the case of behaviours in teamwork and task functions, Li et al. (2001) and Sørensen (2002) affirmed that there often an interplay of rigid social and task behaviours and internal behavioural inconsistencies more in traditional procurement systems than non-traditional procurement strategies.

It is argued that, from both theoretical and practical perspective, cultural and behavioural issues in teamwork and project organizations significantly contribute to role conflict and increase difficulties in the management of the project (Fellow and Hancock 1994; Chan and Tse 2003). To this end, it is affirmed that the unique business environment and behaviours in procurement systems significantly impact on managerial effectiveness, role efficiency and quality of relationship (Xue et al. 2010; Kumaraswamy et al. 2005). Hence, the need to consider environmental and behavioural impact of the attributes of procurement systems is obvious and widely accepted. Thus, a better understanding of influential environmental and cultural factors and behaviours on information sharing and management intuition will help to reduce the adversarial nature of construction procurement systems.

4.4 Implications of These Attributes on Communication Performance

The attributes of traditional and non-traditional procurement systems offer theoretical, practical and structural implications for engendering effective communication performance, communication behaviours, communication networks, communication media and ICT tools in construction project delivery. Increasingly, several studies have argued that to enhance project team communication and information sharing and management, a clear understanding of the communication behaviours and tools that account for the operational business environment in procurement systems must be pursued so as to optimize its benefits and increase performance (Phua and Rowlinson 2003; Cheung et al. 2003; Wong et al. 2005; Diallo and Thuillier 2005; Gorse and Emmitt 2007; Otter and Emmitt 2007). Thus, effective project teams must always adopt a balance of synchronous and asynchronous communication to stimulate effective communication performance (Gorse and Emmitt 2007; Otter and Emmitt 2007).

These attributes herein revealed there may be many behavioural factors such as trust, concern for relationship, incentive, conflict, tension and social networks which may have significant impact on project team communication and effectiveness of their networks (Cheung et al. 2003; Wong et al. 2005; Diallo and Thuillier 2005; Leung et al. 2004). A thorough analysis of such situations can provide a high-level description (role, centrality) through an analysis of relational data since such analysis describes network relations, identifies prominent patterns, traces information flow and measures the strength of network ties in complex communication networks in procurement systems (Garton et al. 1997; Penuel et al. 2006; Xie et al. 2010).

Practically, it can be contended that effectiveness of team communication for the design and delivery of construction project is increasingly becoming crucial due to the growing technical and organizational complexity of construction projects. Hence, adapting information communication technologies (ICTs) to the characteristics of the procurement systems and participants has become a veritable way of improving the performance of design and construction teams (El-Saboni et al. 2009; Lou et al. 2005). This is inherent from the theoretical assertion that ICTs have become synonymous with the better integration of project participants and also a means of improving collaborative working which hitherto was elusive in construction project delivery process (Emmitt and Gorse 2007). This is borne out of the recognition of the need to understand the needs of the individuals and how they communicate in any given procurement system (context) within project teams if communication is to be effective (Emmitt and Gorse 2007).

4.5 Summary

In this chapter, the main typologies of procurement systems in construction project delivery have been briefly elucidated and their inherent attributes and implications for communication remains a key tenets to strategies towards improving communication performance in non-traditional procurement systems. An articulation of the inherent implications of these attributes of procurement systems in construction has been broached. It is expected that a clear understanding of these attributes and their practical implications for communication and information sharing is critical. The next chapter focuses on understanding non-traditional procurement systems, its various types and critical factors in its management.

References

Ashworth, A., & Hogg, K. (2007). *Willis's practice and procedure for quantity surveyor*. Oxford: Blackwell Publishing Ltd.

Assaf, S. A., & Al-Hejji, S. (2006). Causes of delay in large construction projects. *International Journal of Project Management, 24*(4), 349–357.

Beach, R., Webster, M., & Campbell, K. M. (2005). An evaluation of partnership development in the construction industry. *International Journal of Project Management, 23*(8), 611–621.

Bosch-Sijtsema, P., & Postma, T. (2009). Cooperative innovation projects: Capabilities and governance mechanisms. *Journal of Product Innovation Management, 26*(1), 58–70.

Bresnen, M. (2010). Keeping it real? Constituting partnering through boundary objects. *Construction Management & Economics, 28*(6), 615–628.

Chan, D., Chan, A., Lam, P., Lam, E., & Wong, J. (2007). Evaluating guaranteed maximum price and target cost contracting strategies in Hong Kong construction industry. *Journal of Financial Management of Property and Construction, 12*(3), 139–149.

Chan, A. P. C., Chan, D. W., & Yeung, J. F. (2010). *Relational contracting for construction excellence: Principles, practices and case studies*. Abingdon: Spon Press.

Chan, E. H. W., & Tse, R. Y. C. (2003). Cultural considerations in international construction contracts. *Journal of Construction Engineering and Management, 129*(4), 375–381.

Chen, C. J. (2003). The effects of environment and partner characteristics on the choice of alliance forms. *International Journal of Project Management, 21*(2), 115–124.

Cheung, S. O., Ng, T. S. T., Wong, S. P., & Suen, H. C. H. (2003). Behavioral aspects in construction partnering. *International Journal of Project Management, 21*(5), 333–343.

Cicmil, S., & Marshall, D. (2005). Insights into collaboration at the project level: Complexity, social interaction and procurement mechanisms. *Building Research and Information, 33*(6), 523–535.

Diallo, A., & Thuillier, D. (2005). The success of international development projects: Trust and communication: An African perspective. *International Journal of Project management, 23*(3), 237–252.

El-Saboni, M., Aouad, G., & Sabouni, A. (2009). Electronic communication systems effects on the success of construction projects in United Arab Emirates. *Advanced Engineering Informatics, 23,* 130–138.

Emmitt, S., & Gorse, C. A. (2007). *Communication in construction teams*. London: Taylor & Francis.

Eriksson, P. E. (2017). Procurement strategies for enhancing exploration and exploitation in construction projects. *Journal of Financial Management of Property and Construction, 22*(2), 211–230.

Fellow, R. F., & Hancock, R. (1994). Conflict resulting from cultural differentiation: An investigation of the new engineering contract. Council of International Construction Research and Documentation Proceedings, Construction Conflict: Management and Resolution, CIB, Rotterdam, The Netherlands (pp. 259–267).

Garton, L., Haythornthwaite, C., & Wellman, B. (1997). Studying online social networks. *Journal of Computer-mediated Communication, 3*(1).

Gorse, C. A., & Emmitt, S. (2007). Communication behaviour during management and design team meetings: A comparison of group interaction. *Construction Management and Economics, 25,* 1197–1213.

Green, S. D., Fernie, S., & Weller, S. (2005). Making sense of supply chain management: A comparative study of aerospace and construction. *Construction Management and Economics, 23*(6), 579–593.

Hughes, W., Hillebrandt, P., Greenwood, D., & Kwawu, K. (2006). *Procurement in construction industry: The impact and cost alternative markets and supply processes*. London: Taylor and Francis.

Kumaraswamy, M. M., & Dissanayaka, S. M. (2001). Developing a decision support system for building project procurement. *Building and Environment, 36*(3), 337–349.

Kumaraswamy, M. M., Rahman, M. M., Ling, F. Y. Y., & Phng, S. T. (2005). Reconstructing cultures for relational contracting. *Journal of Construction Engineering Management, 131*(10), 1065–1075.

Latham, S. M. (1994). Constructing the team. Joint government/industry review of procurement and contractual arrangements in the UK construction industry. Final Report, HMSO. London.

Lau, J. S. K., Huang, G. Q., & Mak, K. L. (2004). Impact of information sharing on inventory replenishment in divergent supply chains. *International Journal of Production Research, 42*(5), 919–941.

Leung, M. Y., Ng, S. T., & Cheung, S. O. (2004). Measuring construction project participant satisfaction. *Construction Management and Economics, 22*(3), 319–331.

Li, H., Cheng, E. W. L., & Love, P. E. D. (2000). Partnering research in construction. *Engineering, Construction and Architectural Management, 7*(1), 76–99.

Li, J., Lam, K., & Qian, G. M. (2001). Does culture affect behaviour and performance of firms? The case of joint ventures in China. *Journal of International Business Studies, 32*(1), 115–131.

Ling, F., Peng Chong, T., Yan, N., Albert, T., & Asanga, G. (2015). Effect of adoption of relational contracting practices on relationship quality in public projects in Singaporenull. *Engineering, Construction and Architectural Management, 22*(2), 169–189.

Lou, M. M., Liew, P. S., Gu, N., & Ding, L. (2005). An agent approach to supporting collaborative design in 3D virtual worlds. *Automation in Construction, 14*(2), 189–195.

Love, P., Davis, P., Baccarini, D., Wilson, G., & Lopez, R. (2008). Procurement selection in the public sector: Atatle of two state, pp. 1–11.

Love, P. E. D., Skitmore, M., & Earl, G. (1998). Selecting a suitable procurement method for a building project. *Construction Management and Economics, 16*(2), 221–233.

Luu, D. T., Ng, S. T., & Chen, S. E. (2005). Formulating procurement selection criteria through case-based reasoning approach. *ASCE Journal of Computing in Civil Engineering, 19*(3), 269–276.

Manley, K., & Chen, L. (2016). The impact of client characteristics on the time and cost performance of collaborative infrastructure projects. *Engineering, Construction and Architectural Management, 23*(4), 511–532.

Molenaar, K., Sobin, N., Gransberg, D., Tamera McCuen, T. L., Sinem Korkmaz, S., & Horman, M. (2009). *Sustainable, high performance projects and project delivery methods*. The Charles Pankow Foundation and The Design-Build Institute of America.

Morledge, R., & Smith, A. (2013). *Building procurement* (2nd ed.). Oxford: Wiley-Blackwell.

Naoum, S., & Egbu, C. (2015). Critical review of procurement method research in construction journals. *Procedia Economics and Finance, 21*, 6–13.

Ning, Y. (2014). Quantitative effects of drivers nd barriers on networking strategies in public construction projects. *International Journal of Project Management, 23*(1), 75–91.

Otter, A., & Emmitt, S. (2007). Exploring effectiveness of team communication Balancing synchronous and asynchronous communication in design teams. *Engineering, Construction and Architectural Management, 14*(5), 408–419.

Palaneeswaran, E., Kumaraswamy, M., & Ng, T. (2003). Targeting optimum value in public sector projects through 'best value'–focused contractor selection. *Engineering, Construction and Architectural Management, 10*(6), 418–431.

Penuel, W. R., Sussex, W., Korbak, C., & Hoadley, C. (2006). Investigating the potential of using social network analysis in educational evaluation. *American Journal of Evaluation, 27*(4), 437–451.

Phua, F. T. T., & Rowlinson, S. (2003). Cultural differences as an explanatory variable for adversarial attitudes in the construction industry: The case of Hong Kong. *Construction Management and Economics, 21*(7), 777–785.

Rameezdeen, R., & Ratnasabapathy, S. (2006). A multiple decisive factor model for construction procurement system selection. In *Proceedings of the 6th annual research conference of the Royal Institution of Chartered Surveyors*, September, University College London, UK.

Ranganathan, C., & Lerpittayapoom, N. (2002). Towards a conceptual framework for understanding strategic alliance in e-commerce. *Proceedings of the 35th Hawaii International Conference on System Sciences*, Hawaii, January 7–10.

Seeley, I. H. (1997). *Quantity surveying practice* (2nd ed.). London: Macmillan.

Shehu, Z., Endut, I. R., & Akintoye, A. (2014). Factors contributing to project time and hence cost overrun in the Malaysian construction industry. *Journal of Financial Management of Property and Construction, 19*(1), 55–75.

Song, L., Mohamed, Y., & AbouRizk, S. M. (2009). Early contractor involvement in design and its impact on construction schedule performance. *Journal of Management in Engineering, 25*(1), 12–20.

Sørensen, J. B. (2002). The strength of corporate culture and the reliability of firm performance. *Administrative Science Quarterly, 47*(1), 70–91.

Tawiah, P. A., & Russell, A. D. (2008). Assessing infrastructure project innovation potential as a function of procurement mode. *Journal of Management in Engineering, 24*(3), 173–186.

Turner, A. (1997). *Building procurement*. London: Palgrave Macmillan Ltd.

Wong, P. S. P., Cheung, S. O., & Ho, P. K. M. (2005). Contractor as trust initiator in construction partnering—Prisoner's dilemma perspective. *Journal of Construction Engineering Management, 131*(10), 1045–1053.

Xie, C., Wu, D., Luo, J., & Hu, X. (2010). A case study of multi-team communications in construction design under supply chain partnering. *Supply Chain Management: An International Journal, 15*(5), 363–370.

Xue, X., Shen, Q., & Ren, Z. (2010). Critical review of collaborative working in construction projects: Business environment and human behaviors. *Journal of Management in Engineering, 26*(4), 196–208.

Chapter 5
Non-traditional Procurement Systems

Abstract Evidence from the construction industry reveal an urgent need for a change in culture and attitude and transition from fragmentation and adversarial relationship to a more cooperative, integrated and collaborative working relationship. This calls for a drive towards exploring alternative procurement methods as a move away from the prevailing adversarial traditional procurement models. Against this background, various alternative procurement models belong to the group generally referred to as the non-traditional procurement systems. These alternative procurement models are deemed to have benefit of encouraging collaborative working, improved relationships, integrated process and effective cooperation. However, it has been noted that the construction industry has been slow the uptake of these alternative procurement models as well as lacking understanding of how their benefits can be optimized. In spite of their increasing popularity in the global construction industry, an in-depth understanding of the various typologies, their inherent attributes and the critical factors for their performance is lacking. In this chapter, an insight is provided on non-traditional procurement typologies and establishes the critical factors for their success and optimized benefit. In the wake of the changing business environment characterized by tense competitiveness and wide global linkages in the construction industry, this knowledge is timely to facilitate the adoption of the most appropriate alternative procurement towards effective and efficient interorganization collaborative management systems in project delivery.

Keywords Non-traditional procurement · Collaborative working · Efficient inter-organization · Management systems

5.1 Introduction

Within the global construction industry, there is a general consensus that even though non-traditional is thought to offer significant benefits over traditional contracts, a clear understanding of the various typologies and it inherent characteristics is however lacking. Additionally, attempt to gain knowledge and the attributes of the typologies of non-traditional procurement systems is currently disjointed and uncoordinated. The main crux of the evolution of non-traditional procurement systems is to advance

T. E. Kwofie et al., *Effective Construction Project Delivery*,
https://doi.org/10.1007/978-3-030-49374-5_5

towards a philosophy of managing construction project in a more integrated, collaborative manager and enhanced relationship. By gaining an in-depth understanding of the typologies of non-traditional procurement systems, their inherent attributes and antecedent factors for managerial efficiency and effective communication, their benefits can be optimized. The chapter also delineates the various factors in non-traditional procurement systems that have evolved over the years as well as their implications for management, ICT adoption and development and information sharing and management in the global construction industry. This is as a response to the general recognition and growing need to curtail the adversarial and fragmented nature of construction project environment towards a more integrated and collaborative process.

5.2 Understanding Non-traditional Procurement

The main theoretical underpinnings behind the practice and adoption of non-traditional procurement systems are for it to serve as a efficient construction method and organizational approach to analyzing, enhancing and improving construction project delivery process through controlling and managing elements and components of the project and process from inception to completion (Love et al. 2008; Li et al. 2000; Ling et al. 2015). Bennett and Grice (1990) alluded that the urge towards an efficient procurement of a construction project through the choice of the most appropriate procurement strategy has been long recognized a most important determinant of project success. Among numerous practitioners in the global construction industry, there is seemingly a consensus that there is one typology of procurement method that logically appear to be better than all others in some situations for an individual project, but that no one procurement method is likely to be better than others for any project (Love et al. 1998).

Gordon (1994) and Morledge et al. (2006) contended that an appropriate procurement system could reduce construction project costs by an average of 5% as well as enhancing the probability of project success. It is well acknowledged that the selection of an appropriate procurement system could potentially enhance the probability of project success (Naoum and Mustapha 1995; Luu et al. 2005). In practice, some decision-makers often struggle in ascertaining the suitability of various procurement approaches primarily due to the fact that it is virtually impossible for them to capture a diverse continuum of procurement options, client characteristics and needs, project characteristics and external conditions and governance systems through their own experiences of previous projects (Kumaraswamy and Dissanayaka 2001; Morledge et al. 2006; Eriksson 2017; Luu et al. 2005). Non-traditional procurement systems have emerged as a panacea for overcoming the adversarial nature of construction, improving collaboration, ensuring an integrated project delivery process through a number of varied typologies and options.

In essence Li et al. (2000) averred in an attempt to overcome the adversarial nature of construction, partnering and integrated project design (IPD) have consistently been deployed as a strategic mechanism for engendering collaboration between parties so as to attain mutual goals. It has generally been asserted that a clear understanding of procurement systems is based on three principal factors which are governance factors, behavioural factors and business environmental factors (Xue et al. 2010; Morledge et al. 2006; Luu et al. 2005). Of a fact, it is known that non-traditional procurement systems exhibit unique contractual and non-contractual governance systems as well as unique behavioural human factors (Xue et al. 2010; Chen 2003; Cicmil and Marshall 2005). Xue et al. (2010) revealed that most typologies of non-traditional procurement systems have emerged for improving performance and enhancing competitiveness by responding to the changing environment in construction.

For the past two to three decades, it is an established fact that the construction industry has experienced a significant transformation in terms of scale of projects, processes involved in the project execution, complex procurement scenario and more recently advancements in construction technology (Bresnen and Marshall 2000; Lloyd-walker et al. 2014; Donato 2016). Procurement systems have persistently been described as a continuum of various strategies and processes for project organization and management including a selection of the source of funding, partner selection, pricing agreements and allocation of resources and amount of subcontracting (Osipova and Eriksson 2011; Hughes et al. 2015; Perrenoud et al. 2017). Non-traditional procurement on the other hand as an idea of procurement strategy focuses on developing long-term relationships and enhancing productivity that can assist in reducing the potential of cost overrun and hence minimizing disputes (Donato et al. 2015; Ey et al. 2014; Morledge et al. 2006).

Non-traditional procurement systems were deemed integrated procurement routes which simplify procurement from the employer's perspective as they provide a unified interface between the construction project and the employer's organization (Chitkara 2005; Cannon 2006; Belev et al. 2004). Partnering, project alliancing, integrated project delivery (IPD), joint ventures and coalition have been notable forms of non-traditional procurement systems.

5.3 Types of Non-traditional Procurements

Partnering, project alliancing, integrated project delivery (IPD), joint ventures and coalition have all emerged has typologies of non-traditional procurement as an attempt to overcome the adversarial nature of construction, fragmentation in the process and improve relationship (Sha 2016; Xue et al. 2010; Rahman and Kumaraswamy 2012; Molenaar et al. 2010). Additionally, these non-traditional procurement strategies are viewed as alternative way of governing construction projects and have become a historically and contextually embedded practice. No type is fit for all except being considered on project to project and context basis.

5.4 Partnering

Partnering also referred to as partnership from the world of business often denoted a form of business entity in which partners share with each other the profits or losses of the business undertaking in which all have invested (Bennett and Jayes 1995; Bresnen 2010a, b; Dozzi et al. 1996; Larson and Drexler 1997; Eriksson 2010). In construction, partnering refers to an entity which is formed when an entity by an owner, designer and/or contractor with collaborative relationship in an associating process (Bresnen and Marshall 2000, 2010; Beach et al. 2005; Hauck et al. 2004). From the point of view of Bennett and Jayes (1995), partnering is also generally perceived as a management approach used by two or more organizations to achieve specific mutual business objectives and based an agreed method of problem resolution and an active search for continuous improvements.

By comparison with other forms of non-traditional procurement systems such as project alliancing and IPD, it is accounted that the mere definitions of these typologies do not significantly reveal the actual procedural practices or any possible differences between them (Lahdenperä 2012; Wood 2005; Eriksson 2010). The conclusion drawn by Bresnen (2010a, b) suggests that project partnering as a form of integrated and collaborative procurement strategy in any given context is often likely to be a strategic relationship employing a highly specific combination of combination of tools, techniques, processes and practices. This combination of tools, techniques, processes and practices predominantly manifests itself in different ways which makes it hard to generalize it into a universally applicable model (Bresnen 2010a, b; Bresnen and Marshall 2002). Partnering remains perhaps the most significant development till date as a means of improving project performance, while offering direct benefits to procurement and management of construction projects in many countries notably the UK, Australia and New Zealand (Dozzi et al. 1996; Larson and Drexler 1997).

The crux of partnering is to achieve specific business objectives through maximizing the effectiveness of each participant's resources and establishing ongoing business relationships (Bennett and Jayes 1995; Eriksson 2010). Against this background, several studies have suggested that there is no too little doubt about the positive aspects of project partnering arrangements especially in the construction industry. In the illustration demonstrated by Bennett and Jayes (1998), partnering can create a win-win relationships utilizing sophisticated strategy and require a willingness to improve the joint performance which can possibly results in a potential savings of 40–50% in both cost and time.

Ng et al. (2002) also cited instances of success through savings of 40% on costs and 70% on time. Wood and McDermott (1999) and Hamza et al. (1999) averred that partnering is an optimistic and robust approach and move that can indeed lead to benefits for all parties.

From the practical perspective of Lamont (2001), it is suggested that partnering results in people feeling empowered and thereby work together more effectively.

Indeed, there is more ample evidence of successful partnerships which has amplified confidence for many clients and contracting organizations opting for partnering strategy within their commercial relationships. In spite of these positives in partnering, there is also evidence suggesting that in some countries, for example, Australia, the partnering strategy has been over exploited and abused (Broom 2002; Colledge 2005). Wood (2005) and Eriksson (2010) alluded that there has been continuous lack of depth in the analysis and examination of the nature of partnering in practice to justify the claims made in favour of it. Given that most of the focus on partnering has overly concentrated on the success stories, social and psychological issues associated with the successful integration of partnering into an industry which is traditionally adversarial in overlooking which is of such significant importance have been overlooked (Wood 2005; Bresnen and Marshall 2000).

It is worth noting that the past two decades have witnessed a surge in identifying and analyzing the critical factors for partnering relationships success (see Tyler and Matthews 1996; Cheng et al. 2000; Black et al. 2001). Central to these factors have been the emergence of cooperation and teamwork, openness and honesty, trust, equity and equality and a transparent communication and information sharing system (Green 1999; Taylor 1999; Ng et al. 2002; Tyler and Matthews 1996; Cheng et al. 2000; Black et al. 2001). Additionally, other stakeholders and commentators place considerable emphasis upon the changing attitudes, improving interpersonal relationships and transforming organizational cultures as important elements for success (Bresnen and Marshall 2000).

5.5 Project Alliancing

Project alliancing is described as a method of managing and delivery major capital assets such as project with the main focus of the owner and participants working together as an integrated collaborative team acting in good faith, with integrity and making unanimous best-for-project decisions, managing all risks of project delivery jointly and sharing the outcome of the project (Hauck et al. 2004; Lahdenperä 2012). Abrahams and Cullen (1998) also referred to project alliancing as a form of agreement between two or more entities who agree to cooperatively work together on the basis of sharing risk and reward aimed for the purpose of achieving agreed outcomes based on principles of good faith and trust. Often to many, it is difficult to draw the distinction between either project partnering and project alliancing or with other forms of non-traditional procurement. Hauck et al. (2004) intimated that in project alliancing, risk allocation and remuneration which are the hard contractual issues that affect entities' bottom lines are predominantly used. This is regularly perceived as the obvious difference between project alliance and partnering (Hauck et al. 2004).

In a more simple terms and focusing on the construction industry, project alliance refers to where a building owner forms a collaborative alliance with one or more service providers or organizations such as designers (consultants), contractors and sub-contractors for the pure purpose of delivering an outstanding results of specific project by sharing risks (Laan et al. 2011; Rahman and Kumaraswamy 2004). It is generally an accepted fact that project alliancing traces its roots from industry rather than construction as its early application was witnessed in the early 1990s in the petroleum sector as a collaborative process for an oil project in the North Sea in the UK by the British Petroleum (BP) (Knott 1996). After BP's initial unsuccessful approach to the oil project in the North Sea in the UK using traditional approach to engineering, a cheaper approach through the use of latest technology at the time was also explored using project alliancing procurement model (Knott 1996; Sakal 2005). With the traditional approach yielding minimal results by way of cost savings, it became obvious that there is the need for a paradigm shift in the approach to a more drastic strategy (Barlow 2000; Walker and Hampson 2003). As a result, BP explored alliance agreements with well-defined scopes which was separate from the main contracts and pursued behaviour change as an essential factor for success in alliancing (Barlow 2000; Walker and Hampson 2003).

The positive experiences and success drawn by BP from the alliancing arrangements led to countries such as Australia adopting the model in oil and gas projects in 1994 (Sakal 2005; Morwood et al. 2008). Through continued diffusion of awareness, the alliance model was widely adopted soon thereafter on the continent with the first application of project alliance in construction starting in 1997 (Manley 2002; Rowlinson et al. 2006; Morwood et al. 2008). With the account given by Manley (2002) and Hellard (2002), witnessing the active use of PA significantly contributed to the favourable attitude towards PA in Australia, and subsequently, it launches in the UK. From the initial implementation PA in the petroleum industry, there was a general realization that apart from a drastic change in behaviour, the PA as a new contracting strategy would need to create an environment that necessitated commitment of the teamwork, relationship development and trust (Lahdenperä 2012; Sakal 2005; Hellard 2002; Tyler and Matthews 1996).

From the construction industry, it is a fact that the few projects applying PA were realized in the late 1990s, and the flow of PA projects was also continuous; the actual significant breakthrough occurred from the mid 2000s (Lahdenperä 2012; Sakal 2005). So far, numerous projects have been implemented by the PA system while the arrangement has developed substantially and is based on a multiparty contract with joint liability (Sakal 2005). These projects have largely been roads, rail, water and power infrastructure projects. Hauck et al. (2004) asserted that project alliance practice predominantly evolved from the need to improve the implementation of complex, demanding and risky investment projects. The tenets of PA contracting and its inherent attributes resulted in a paradigm shift in the traditional practice of contractors making suboptimal project decisions in order to optimize or protect their profit or interest (Walker and Hampson 2003; Hauck et al. 2004). PA approach has consistently adopted a multiparty contract for a while partial-price team selection, non-price selection and occasionally, full price selection have emerged as other recent reforms (Rowlinson et al. 2006; Greenham 2007).

Ross (2006) revealed that generally, certain elements in project alliancing will differ from one project to the other to optimize the probability for success; all project alliance projects will ostensibly exhibit the these attributes and characteristics. These are sharing of all uninsurable risks by all participants, using a three-limb and open book compensation model for paying all participants, adopting project alliance board as a governance system, project management team made up of all participants and resolving alliance disputes internally with litigation being last resort (Sakal 2005; Ross 2006).

5.6 Integrated Project Delivery (IPD)

The definition of IPD provided by the American Institute of Architects (AIA) has remained the widely adopted one in several research and scholarships. The AIA defines IPD as "a project delivery approach that integrates people, systems, business structures, and practices into a process that collaboratively harnesses the talents and insights of all project participants to optimize project results, increase value to the owner, reduce waste, and maximize efficiency through all phases of design, fabrication, and construction" (AIA California Council 2007, p. 6). IPD is deemed as a relatively recent addition to the building process practice and seems to have emerged initially with that name in 2003 (Matthews and Howell 2005; Perlberg 2009; Sive 2009).

Initially, the IPD arrangement was deemed to be closer to design–build contract for construction of utilities and it also involved adhering to the early involvement practice so that the price was set later and the companies were to share the financial gain or loss inherent from the project (Lahdenperä 2017; Matthews and Howell 2005; Ren et al. 2006). Kent and Becerik-Gerber (2010) alluded that integrated project delivery (IPD) seeks to improve project outcomes through a collaborative approach of aligning the incentives and goals of the project team through shared risk and reward, early involvement of all parties, and a multiparty agreement. According to Xue et al. (2010), IPD has a primary focus of bridging the separated construction life cycle stages, activities, teams or information flow.

From extant literature, it is clear to not that early involvement and integration of versatile expertise, systems and business practices remain the crux of IPD (Lee et al. 2012; AIA 2010; AIA 2007). From this, the AIA and Associated General Contractors of America (AGC) have remained at the centre of establishing standards, disseminating guidelines and facilitating the practice and adoption of IPD (AIA 2009). The works of Perlberg (2009), Parrish et al. (2008) and Lichtig (2005, 2006) revealed that still IPD continues to be continues to be a multiparty IPD agreement as well as created the contractual and financial framework to facilitate the effective collaboration between construction project participants. However, in spite of these efforts, there is yet to be a standard definition of IPD that has been accepted by the industry at large. There are still incidence of different definitions and widely varying approaches and sophistication levels (Sive 2009; Parrish et al. 2008). This suggests that the term IPD denotes and describes significantly varying contract arrangement and team processes (Sive 2009; Perlberg 2009; AIA 2010).

It can also be said that in spite of these notable discrepancies and seemingly lack of consensus on IPD, literature is also replete with consistent similarities that have been found within most IPD projects and definitions. Consistently, most definitions and principles of IPD from various projects have entailed multiparty agreement, early involvement of all parties and shared risk and reward (AIA 2007, 2010; Middlebrooks 2008; Lichtig 2005; Parrish et al. 2008; Kent and Becerik-Gerber 2010). However, Kent and Becerik-Gerber (2010) contended that the general acceptance is that not all of these principles are necessary in order to constitute IPD as there are other principles that may be regarded as important for project success. However, the three named principles are well supported due to their integral nature and inclusion in the majority of IPD projects and extant literature supporting IPD. From a very comprehensive overview and outlook, it can be stated that IPD practice seems to merge and exploit beneficial features and ideas from various sources and earlier experiences (Kent and Becerik-Gerber 2010).

5.7 Joint Ventures

There is enough evidence from existing body of knowledge indicating that joint venture has been described in many ways with many elements and purposes. Simply, the term joint venture is often defined as the "commercial agreement" between two or more companies with the primary motive of allowing greater ease of work and cooperation towards achieving a mutual goal, through the manipulation of the appropriate resources (Kling and Burley 1991; Norwood and Mansfield 1999; Sillars and Kangari 2004). Norwood and Mansfield (1999) explained joint ventures as a scenario where two or more persons or companies combine property, money, skill, knowledge, land, capital, intellectual property, credentials, equipment, etc. to carry out a single business enterprise or a series of business enterprises for profit. In joint venture arrangements, the contribution resources (i.e. property, money, skill, knowledge, land, capital, intellectual property, credentials, equipment, etc.) are normally done equally or in specific proportions such as percentages (Sillars and Kangari 2004; Norwood and Mansfield 1999). However, some of the resources such as skills and knowledge often suffer accurate measurement (Norwood and Mansfield 1999).

Aimin and Barbara (2001) defined construction joint venture as a form of partnership of two or more contractors who form a business alliance with the sole purpose of undertaking a project with shared risks and benefits. From many perspectives, joint ventures have been seen as a form of commercial alliance founded between two or more separate entities or business units by sharing risks and rewards in activities underaten. However, it is generally asserted that joint ventures as a form of non-traditional procurement strategy seem to offer a little more stability in what is often a volatile and "short-term project" construction industry (Sillars and Kangari 2004; Norwood and Mansfield 1999; Bygballe et al. 2015). In the opinion of Sillars and Kangari (2004) and Norwood and Mansfield (1999), joint ventures can be relatively straightforward in setting up and bring to a close without the need to form a separate legally incorporated company. But Norwood and Mansfield (1999) consider joint ventures as a special form of alliance or partnership but temporary entity in nature for the purpose of pursing and performing a project.

Against this background, it has been observed that joint venture arrangements substantially apply same rules which are applicable partnership and alliance (Sillars and Kangari 2004; Norwood and Mansfield 1999). In a typical joint venture, other members' action can bind other associates by a contract which is in furtherance of the enterprise or within the scope of activity (Kling and Burley 1991; Norwood and Mansfield 1999). Thus, each member in the joint venture often called the venturer is liable for the performance of the entire contract and the payment of all labour, material, equipment and other obligations (Bygballe et al. 2015).

5.8 Strategic Alliance

Strategic alliances as a business model have increasingly become a major module of business entities' global strategies explored to achieve various strategic goals (Judge and Dooley 2006; Mockler 2001; Patel 2007). With increased levels of extensive global competition and internationalization coupled with huge complexities of construction project and demanding construction process, construction industry is deemed to benefit from strategic alliances. It is also asserted that alliance partners are able to offer greater knowledge and skills that will position the alliance in a stronger competitive position (Xue et al. 2010; Kwok et al. 2000; Love and Gunasekarn 1999). Strategic alliance has extensively been explored in the field of marketing, pharmaceuticals, manufacturing, aviation and banking industries (see Anand and Khanna 2000; Andersen and Katz 1998; Chen and Popovich 2003; Cravens et al. 2002).

Cravens et al. (2002) defined strategic alliance as a voluntary business relationship formed by two or more organizations that share (proprietary), participate in joint investments and develop linked and common processes to increase the performance of both companies. Hauck et al. (2004) also defined strategic alliance as a form of an interorganizational arrangement usually between two companies that extends beyond a specific project. According to Kwok et al. (2000), the main motive for many organizations forming strategic alliance is to increase the performance of their common supply chain. Chen and Popovich (2003) contended that strategic alliance is solely depended on trust and faith in relationships across various stages of the supply chain to engender success and achieve objectives. Hauck et al. (2004) revealed that the construction industry can significantly benefit from alliances that offer alliance partners to be able to provide greater knowledge that will place the alliance in a stronger competitive position when compared to the situation of the individual company. However, given that businesses are often complex entities that require significant resources and time to develop effectively, entities are sometimes challenged to enter new markets, fill strategic gaps, and increase technological capabilities (Hauck et al. 2004; Kwok et al. 2000).

Strategic alliance is sometimes actually perceived as a collaboration of firms to work together to form a greater effect than before with the motive of improving performance which are seen in decision-making, easy coordination, increase in supply chain productivity and ensuring proper sharing of sales and production information (Li et al. 2000; Andersen and Katz 1998; Chen and Popovich 2003; Cravens et al. 2002). According to Wheelen and Hungar (2000), access to technology, access to specific markets, reducing financial and political risk, achieving or ensuring competitive advantage have been some of the recent motives driving strategic alliances. Despite these, Pett and Dibrell (2001) recounted that shrinking world markets and increased competition have become dominant and compelling reason for recent growth in the number of strategic alliances.

It is worthy to note that, in spite of numerous positive accounts on strategic alliance across various industry typologies, inter-firm linkages have been frequently fraught by of instability, poor performance and termination problems (Parkhe 1993). In overcoming these challenges, existing literature is replete with strategic alliance motives, environment (context), asset specify and perception of opportunistic behaviour, partner selection, trust, commitment and communication between supply chain partners and culture factors critical towards alliance successes (Kaplan et al. 2001; Ranganathan and Lerpittayapoom 2002; Yasuda 2005; Young-Ybarra and Wiersema 1999; Judge and Dooley 2006; Kwon and Suh 2005; Robson 2002; Van Vijfeijken et al. 2002; Elmuty and Kathawala 2001; Bamford et al. 2003; Beugelsdijk et al. 2006).

5.9 Coalition in Construction Supply Chain

Coalition as a business operational and management model is seen as one of the emerging results of innovations in procurement and project management approaches which to some scholars is one of the transitions in the construction industry. Coalition in construction supply chain is also regarded as a form of alliance among entities which allows for cooperation in a joint action, each in their own interest (Xue et al. 2010; Winch 2010; Pryke and Pearson 2006; Pryke 2004, 2005). This form of alliance may be temporary or a matter of convenience. Pryke (2004) conteneded that coalition as a new strategy reguires and indepth analytical method that deals with coalition partner or actor inter-dependence and provides an appropriate level of detail and information in relation to the dynamic, complex, iterative and interactive process often encountered in construction project delivery.

Construction coalition traces its roots from economics which encapsulate maximizing mutual trust between coalition actors for increased profit among group of companies (coalators) (Pryke 2005; Winch 2010; Pryke 2004). In the considered opinion of Pryke (2005), intracoalition in construction projects can be represented as a multi-layer of interdependent networks such as contractual relationships, information exchange and performance incentives.

5.10 Relational Contracting

Sha (2016) opined that relational contracting (RC) as an alternative organizational and governance approach to construction project delivery can well be understood as a historically and contextually embedded practice. From a law scholars' approach and new institutional economics approach to relational governance and contracting, relational contracting as practiced in construction project delivery is conceptualized as "an intentionally incomplete, largely self-enforcing arrangement that places emphasis on renegotiation, realignment of the interests of contracting parties and restoration of the efficiency of the project" (Rahman and Kumaraswamy 2002, p. 12). Relational contracting emerged as a response to the pursuit of a more collaborative working and a move away from a "pure" project management to a less adversarial project governance that respond to emerging challenges arising out to intense competitive globalization and changes in government policies, legal frameworks in project delivery, the extension of private sector involvement, the trend of servitization of construction process (Hartmann et al. 2010; Gottlieb and Jensen 2012; Henisz et al. 2012).

Inherent from the urge towards a more integrated, collaborative working and improved relationship in construction project delivery, relational contracting has increased at a frantic pace, and thus, cooperative relationships in construction have become increasingly common over the past decades (Poppo and Zenger 2002; Lu et al. 2015; Chan et al. 2010).

This has spurred a subsequent stream of research exploring various forms of RC, practices, theoretical positions of relational contracting and factors critical for success (see Tennant and Fernie 2012; Jeffries and Reed 2000; Poppo and Zenger 2002; Lu et al. 2015; Henisz et al. 2012; Rahman and Kumaraswamy 2002; Rahman et al. 2007; Chan et al. 2010; Dewulf and Kadefors 2012; Gottlieb and Jensen 2012).

From a synthesis of extant literature and existing knowledge, relational contracting is not a standalone procurement system but rather a form of organizational and operational model of procuring and managing projects which take the forms of project partnering, strategic partnering, project alliancing, strategic alliancing, integrated project delivery, public–private partnerships and joint venture (Tennant and Fernie 2012; Xue et al. 2010; Pryke and Pearson 2006; Chan et al. 2010; Jacobsson and Roth 2014; Yeung et al. 2012). Among the areas which have been the focus of research in relational contracting are exploring the drivers and barriers to adopting RC practices (Rahman and Kumaraswamy 2002; Rahman et al. 2007; Chan et al. 2010; Dewulf and Kadefors 2012; Gottlieb and Jensen 2012) causal effects between RC practices and attainment of critical success factors (Pryke and Pearson 2006; Doloi 2009; Suprapto et al. 2015; Ning and Ling 2015) and a comparative studies between conventional contractual arrangement and relational governance in construction procurement systems (Poppo and Zenger 2002; Lu et al. 2015; Bygballe et al. 2015).

It is crucial to note that, in spite of the seemingly significant advances and contributions made by way of research and contribution to knowledge, the area of relational contracting especially in construction still has fledgling field that has the potential to make a major scholarly and practical contribution. These areas are the conceptualization of RC, the governance systems in RC and lack of generalization of some findings due to limited sample spaces (Chan et al. 2010; Lahdenperä 2012; Henisz et al. 2012; Jacobsson and Roth 2014; Winch 2010). From extant literature, relational contracting is also conceptualized as a form of collaborative working (see Xue et al. 2010; Winch 2010; Sha 2016; Tennant and Fernie 2012; Brown et al. 2004; Jeffries and Reed 2000). Thus Sha (2016) and Brown et al. (2004) revealed that quasi RC such as design–bid–build (DBB) and design–build (DB) procurement strategies, and "true" RC such as project partnering and alliance agreements are the two main clusters of relational contracting employed in construction project delivery. It must be noted that the explanation of these two types is beyond the scope of this book and also the communication dimension on of non-traditional procurements shall be on the true RCs.

5.11 Critical Factors in Non-traditional Procurement Systems

Existing theoretical and empirical works have revealed and validated that the effectiveness and success of non-traditional procurement systems in managing, governing and operational functions of construction projects are a function of various factors that relate to either the projects attributes and characteristics and enabling factors that could be structural, organizational culture, technological, human behaviours, governance and business environment (context) (Xue et al. 2010; Chen 2003; Cicmil and Marshall 2005; Fong and Lung 2007; Bresnen and Marshall 2000; Beach et al. 2005; Hauck et al. 2004; Drexler and Larson 2000; Dozzi et al. 1996; Anvuur and Kumaraswamy 2007; Ingirige and Sexton 2006).

The need to consider environmental impact, human behaviours, governances systems, culture and technological innovations on management process and operational task functions especially in various procurement systems has become widely accepted (Chen 2003; Sha 2016; Tennant and Fernie 2012; Lahdenperä 2012; Hartmann and Bresnen 2011; Henisz et al. 2012; Gottlieb and Jensen 2012; Pryke 2005).

5.12 Construction Business Environment

The construction business environment (context) is said to exhibit incidence of a lack of genuine cooperation over time, complexities with heightened fragmentation and adversarial relationship and temporary nature (Cicmil and Marshall 2005; Henisz et al. 2012; Suprapto et al. 2015; Sha 2016; Henisz and Zelner 2005). Xue et al. (2010)

alluded that collaborative business culture, attitudes and relationship focused working forms are emerging in the construction industry as a solution to these attributes of the industry. Cicmil and Marshall (2005) noted that construction project environment generally shows a complexity process and varied social setting which have significant implications for trust and relationship building.

Thus, a seamless collaboration and integration across all the project life cycle phases is critical (Ngowi and Pienaar 2005; Green et al. 2005). It is however noted that significant research on the business environment in construction has paid much attention to the construction phase. The argument in favour of a synchronized seamless synergy across all phases is that, practically, all the life cycle phases are not mutually exclusive, and thus, a lack of effectiveness at any of the phases can potentially affect the other directly or indirectly. Given the trends in macro-market globalization and micro-changes in the industry as well as management of firms and organizations, an urge towards a more collaborative and integrated business environment in construction is an imminent necessity that must be pursued rigorously.

5.13 Incidence of Cultural Diversity and Variabilities

The influence of culture on managerial efficiency, team effectiveness and project performance is increasingly becoming an important subject given that these are important factors that have immense impact in establishing sound alliancing approach to projects (Phua and Rowlinson 2003). Organizational norms, culture and practices are said to vary from one tem to other, one organization to the other and from one content to the other (Xue et al. 2010; Ravasi and Schultz 2006; Li et al. 2001; Sørensen 2002). Loosemore (1999) teams, organization and firms develop strong attachment to culture and norms which are often an embedded shared values, understandings, assumptions and goals learned from earlier generations. This consequently results in common attitudes codes of conduct and expectations that guide behaviour making it almost near impossible to yield to other cultures from other organizations or teams (Ravasi and Schultz 2006; Loosemore 1999).

Organizational or team culture becomes a set of shared mental assumptions that guide interpretation and action in organizations by defining appropriate behaviour for various situations and serves as a means to accomplish things (Ravasi and Schultz 2006; Stock et al. 2007). Various scholarly submissions claim that strong organizational and team cultures improve a firm's performance by facilitating internal behavioural consistency (Li et al. 2001; Sørensen 2002). However, this strong team or organizational culture can equally have strong negative impact on interorganization effectiveness and team efficiency (Chan and Tse 2003; Brown and Starkey 1994). Fellow and Hancock (1994) revealed that cultural issues present a significant contribution to conflict among teams, parties and organizations and increase difficulties in the management of projects especially in internationalized projects. Significant interorganizational and intraorganizational and team cultural differences can pose a challenge to effective management and effectiveness in project delivery

(Chan and Tse 2003). Against this, it can theoretically stress that an understanding of the significant cultural factors in teamwork and organizational task function can be a good recipe in ameliorating conflicts or reducing them to manageable levels (Sørensen 2002; Chan and Tse 2003; Brown and Starke 1994). Kwan and Ofori (2001) contended that by a recognition of the impact of culture and cultural factors, a sensible manipulation of the cultural differences can induce positive improvements in the team dynamics, managerial effectiveness and efficiency in project delivery and performance. This could also potentially facilitate partnering and alliance adoption and implementation (Phua and Rowlinson 2003; Kwan and Ofori 2001).

Given that construction delivery process is a varied social setting (Cicmil and Marshall 2005), understanding the cultural orientation of teams and organizations can also trigger the needed cooperative behaviour that can facilitate bonding and teamwork (Phua and Rowlinson 2003). The theoretical foundation for this is that strong organizational and team cultures have high propensity to facilitate a stable environment that can generate reliable performance (Sørensen 2002; Phua and Rowlinson 2003). In contrast, the opinion of Sørensen (2002) is that a volatile environment inherent from a weak cultural orientation among teams and organizations often results in the benefits of being drastically attenuated. Hence, strong cultural teams and organizations though most of the time excel at exploiting established competencies, they often have difficulty exploring and discovering new competencies that better suit changing environmental conditions (Sørensen 2002). According to Rao and Swaminathan (1995) culture from both the micro and macro level have significant impact strategic alliances.

Rao and Swaminathan (1995) revealed that naturally, organizational culture could be strategic, contextual, prevailing norm and mindset, procedures within the managerial functioning of the firm or teams. Hence, developing a sense of openness, mutual respect, trust, and strategic complementarities among strategic partners and alliances can trigger success even in situations of dissimilar but compatible cultures.

5.14 Governance Systems in Construction Project Delivery

A revelation provided by Chen et al. (2018) indicates that in spite of the increased adoption of collaborative working in infrastructure project delivery globally due to their potential to deliver improved project performance over traditional models, project outcomes continue to be unpredictable. Choosing a procurement topology to deliver a given project has remained a major choice that clients make but literature is filled with significant evidence alluding that governance choices for project execution are more important but often overlooked by clients (Chen et al. 2018; Lahdenperä 2012; Zimina et al. 2012; Abdi and Aulakh 2017). Chen et al. (2018) established that clients invariably tend to focus more on procurement choices and pay less attention to differentiating governance systems based on those choices.

It is a fact that optimal governance configurations vary on the basis of the chosen type of team configuration, organizational disposition or procurement strategy (Chen et al. 2018; Chen and Manley 2014; Davis and Walker 2009). Hence, a need to now orient the focus of client to governance over procurement strategy in infrastructure project delivery is a necessary imminent change. Chen et al. (2018) explained governance as the metaframework that guides and controls decision-making on projects. Chen et al. (2018) and Chen and Manley (2014) validated governance structures in relational and collaborative infrastructure projects and revealed contractual and non-contractual governance mechanism that influence project performance in respect of time and cost, innovation and collaboration and sustainability. They reckon that varied forms of these governance mechanisms influence single team or multiple team approach to relational or collaborative infrastructure delivery.

The contractual governance mechanisms entailed risk and reward sharing regime and responsibilities, whereas leadership, team workshops, communication systems, relationship management, design integration constituted the non-contractual mechanisms (Chen et al. 2018; Chen and Manley 2014; Love et al. 2010; Walker and Rahamani 2016). Practically, clients chose from among a range of governance mechanisms to ensure fair risk and reward sharing regimes and actions based on leadership and workshops to build trust and an integrated project team (Abdi and Aulakh 2017). It is generally accounted that in many typical construction project management genre to project delivery, governance structures remain the core tenets through which the intrinsic attributes and managerial characteristics of the project are managed (Lahdenperä 2012; Zimina et al. 2012; Abdi and Aulakh 2017). This is achieved through the development and application of sophisticated collaborative governance structures of both contractual and non-contractual governance mechanisms (Lahdenperä 2012; Zimina et al. 2012; Abdi and Aulakh 2017). Hence, without an efficient governance structures and mechanism, it is difficult to engender success in collaborative relational infrastructure delivery.

Thus, from extant literature and scholarly theoretical positions in both general management and construction project management, the need for a carefully balanced contractual and noncontractual governance mechanisms for effective interbusiness collaboration is perceived as the veritable way to engender effective governance in relational collaborative infrastructure delivery (Reuer and Devarakonda 2016; Salvato et al. 2017; Krishnan et al. 2016). Some studies have affirmed that in many alliances, contractual mechanisms suited property-based assets, whereas non-contractual governance was found to be best suited to knowledge-based assets (Hoetker and Mellewigt 2009; Reuer and Devarakonda 2016). Subsequently, Hoetker and Mellewigt (2009) further affirmed that the two main types of governance are not mutually exclusive, and accordingly, a variegated governance of some form is optimal.

5.15 Human Behaviours in Project Delivery

Redefining project delivery process and environment as a social interaction endeavour in a complex network and organizational process perspective has been deemed as an important direction in human behavioural factors in project management and procurement systems (Wong et al. 2005; Cheung et al. 2003; Turner and Müller 2003). However, in spite of the contribution of behaviour to the effectiveness of collaborative and relational contracting and working models in construction project delivery, continuous analysis has remained relatively small (Xue et al. 2010; Phua and Rowlinson 2003; Cheung et al. 2003).

Progressively, several researches have made compelling argument in favour of continuous studying of the attitudes and behaviours of individuals involved in the project environment (Phua and Rowlinson 2003; Cheung et al. 2003; Wong et al. 2005; Turner and Müller 2003). Literature is replete with several and varied behavioural factors which are of significance to alliances, partnering and relational models of project delivery. Li et al. (2009) contend that human behavioural factors are important for team effectiveness, organizational efficiency and improving relationship in project delivery. According to Ibrahim et al. (2019), varied forms of information sharing behaviours have been compelling issues that have stifled the success of ICT adoption in the construction industry especially in the area of project team communication management. In collaborative and integrated working models which is identical to most non-traditional procurement systems, trust, concern for relationship, incentive, conflict, tension have become notable human behavioural factors in construction project delivery (Cheung et al. 2003; Wong et al. 2005; Diallo and Thuillier 2005; Leung et al. 2004).

Trust has remained a dominant and popular human behaviour factors critical for success in collaborative and relational working models and considered as a multi-level phenomenon that exists at the personal, interpersonal, organizational, interorganizational and significantly at international levels (Das and Teng 2001; Cheung et al. 2003). Das and Teng (2001) posited that trust is a key behaviour that helps in building cooperative relationships in working models. Cheung et al. (2003), on the other hand, deemed trust building as the central attitudinal factor in multi-team level working and indispensable to partnering arrangement. Accounts by Diallo and Thuillier (2005) and Rahman and Kumaraswamy (2005) reveal that right communication behaviours can lead to trust building and cooperation in collaborative working models. By way of competency behaviours, Wong et al. (2005) identified performance (partner's competence) and permeability (partner's openness in sharing information) as critical to team effectiveness and success. Vaaland (2004) opined that collaboration among partners and alliances can be significantly improved in tension-induced relationship by adopting a more relaxed view about the role of conflict. This tends to create a more replaced and non-adversarial project environment.

It is also suggested that instituting appropriate incentives in project delivery are deemed as a good behaviour that can contribute to effective project governance

(Pryke and Pearson 2006; Pryke 2004, 2005). Others have also consistently itemized behavioural management mechanism, relationship management and confidence building as behavioural measures that have high propensity to foster effective collaborative and integrated alliances, partnerings and coalitions in construction project environment (Harmon 2003; Vaaland 2004; Leung et al. 2004). They are deemed as salient non-quantifiable behaviours that can impact a project environment positively or negatively.

5.16 Information and Communication Technology (ICT) Effects in Construction

Information and communication technology (ICT) is perceived to revolutionize the construction industry due to its anticipated impact of radically improving collaboration among the wide range of tasks and functions in construction life cycle phases and to increase efficiency (Holzer et al. 2007). This perception was borne out of a cue from the revolution ICT adoption created in other industries such as manufacturing, production, marketing and services. Given the digitalized era the industry finds itself, ICT tools come in handy in managing the huge volume of information needed and processed in the industry in the delivery process ensure positive project performance (Egbu and Robinson 2005).

Bouchlaghem et al. (2005) proved that computer-aided techniques and tools which are forms of ICT in construction can facilitate collaborative and relational working among geographically distributed and colocated project team participants across the project life cycle phases. It can be asserted that collaboration and integration remain the crux of non-traditional procurement which potentially can remedy the adversarial nature of traditional procurement systems. From this, Otter (2005) and Mead (1999) revealed that computer-aided techniques and tools are necessary for facilitating collaborative infrastructure project. From the last few decades, the construction industry has explored new ICT tools that are aimed at enhancing distributed organizational interactions and achieve good coordination and effective communication between distributed project teams (Perry and Sanderson 1998; Wikforss and Lofgren 2007). The development of Lou et al. (2005) was a multi-agent system that supports collaborative design in 3D virtual environment as well as facilitating the level of communication not readily available in conventional computer-aided design (CAD) systems. Similarly, Duan and Zhou (2006) proved that the use of computer-supported collaborative design (CSCD) system can ensure the coherence of product information in the design process through integrating CAD system with product data management (PDM) system.

In the area of design and management, improving accessibility and flexibility in collaborative design was facilitated by real-time online collaborative computer-aided design (Chen and Tien 2007). In general, it is a common knowledge that information, computing and communication technologies adopted in construction project delivery have contributed to an important role in improving design communication (Demirkan 2005). Nonetheless evidence gathered from Perry and Sanderson (1998), and Wikforss and Lofgren (2007) suggest that sometimes, some of the ICT innovations and tools adopted have not yielded the necessary impact.

Today, more advanced and tailored information technologies (IT) and tools are changing the way the world manage construction information and communication among multi-level teams. The growth of web-based tools and technologies offers construction organizations a means to optimize project communications. Mead (1999) and Otter (2005) intimated that the development of intranets, internets and web-based sites for projects can enhance project communication by granting team access to a common, centralized database of construction information by all project participants. The COMPAS by the construction industry institute (CII) has contributed immensely to evaluating communication performance in project teams (Thomas et al. 1999). This has been used by various researchers in evaluating communication challenges in a social network analysis (see Xie et al. 2010, 2000; Dawood et al. 2002). Xie et al. (2010) found that partnering and alliances can be enhanced through the use of ICT web-based tools. From both theoretical and practical inference, it can be stated that the construction industry has witnessed an array of ICT tools which when used properly, project and context specific system can be a useful actor in a project communication network and consequently influence information timeliness, access, understanding, procedure and protocols. However, information overload, underload and gatekeeping can also be notable negative effects of ICT tools especially in intranets.

5.17 Summary

This chapter has given a conceptual understanding of non-traditional procurement system and elucidated the various typologies employed in construction project delivery. Various critical factors in non-traditional procurement systems that have significant impact of the success and effectiveness in practice were also discussed. By this a clear understanding of non-traditional procurement typologies, the chapter has further the evolution of the concept from its application in other industry. Even though, the various typologies seem similar and generates confusion sometimes, a clear attempt has been made to present the unique intrinsic traits and attributes of the various typologies.

References

Abdi, M., & Aulakh, P. (2017). Locus of uncertainty and the relationship between contractual and relational governance in cross-border interfirm relationships. *Journal of Management, 43*(3), 771–803.

Abrahams, A., & Cullen, A. (1998). Project alliances in the construction industry. *Australia Construction Law Newsletter, 62,* 31–36.

AIA and AIACC. (2007). *Integrated project delivery: A guide.* Retrieved from http://info.aia.org/SiteObjects/files/IPD_Guide_2007.pdf (11/02/2020).

AIA. (2009). *Lessons learned from applied integrated project delivery.* Retrieved from http://webcache.googleusercontent.com/search?q=cache:cniTjp8TMFYJ:www.ipd (11/02/2020).

AIA. (2010). *Integrated project delivery: Case studies.* Sacramento, CA: American Institute of Architects California Council.

AIA California Council. (2007). *Integrated project delivery: A working definition.* http://www.ipdca.net/images/Integrated%20Project%20Delivery%20Definition.pdf. January 22, 2020.

Aimin, Y., & Barbara, G. (2001). Negotiating control and achieving performance in international joint ventures: A conceptual model. *Journal of International Management, 7,* 295–315.

Anand, B., & Khanna, T. (2000). How entrepreneurial firms can benefit from alliances with large partners. *Academy of Management Executive, 15,* 139–148.

Andersen, M. G., & Katz, R. B. (1998). Strategic sourcing. *The International Journal of Logistics Management, 9*(1), 1–14.

Anvuur, A. M., & Kumaraswamy, M. M. (2007). Conceptual model of partnering and alliancing. *Journal of Construction Engineering and Management, 133*(3), 225–234.

Bamford, J. D., Gomes-Casseres, B., & Robinson, M. S. (2003). *Mastering alliance structure: A comprehensive guide to design, measure, and organization.* San Francisco, CA: Jossey-Bass.

Barlow, J. (2000). Innovation and learning in complex offshore construction projects. *Research Policy, 29,* 973–989.

Beach, R., Webster, M., & Campbell, K. M. (2005). An evaluation of partnership development in the construction industry. *International Journal of Project Management, 23*(8), 611–621.

Belev, G., Berg, A. C., & Doll, A. J. (2004). *A guide to the project management body of knowledge* (pp. 269–297), 3rd edn, Project management institute, four campus boulevard, Newtown Square, Pennsylvania, USA.

Bennett, J., & Grice, A. (1990). Procurement systems for building. In P. S. Brandon (Ed.), *Quantity surveying techniques: New directions* (pp. 243–262). BSP Professional Books.

Bennett, J., & Jayes, S. (1995). *Trusting the team: The best practice guide to partnering in construction.* Centre for Strategic Studies in Construction/Reading Construction Forum, Reading.

Bennett, J., & Jayes, S. (1998). *The seven pillars of partnering.* Reading Construction Forum, Reading.

Beugelsdijk, S., Koen, C. I., & Noorderhaven, N. G. (2006). Organizational culture and relationship skills. *Organization Studies, 27*(6), 833–854.

Black, C., Akintoye, A., & Fitzgerald, E. (2001). An analysis of success factors and benefits of partnering in construction. *International Journal of Project Management, 18,* 423–434.

Bouchlaghem, D., Shang, H., Anumba, C. J., Cen, M., Miles, J., & Taylor, M. (2005). ICT-enabled collaborative working environment for concurrent conceptual design. *Architectural Engineering and Design Management, 1*(4), 261–280.

Bresnen, M. (2010a). Keeping it real? Constituting partnering through boundary objects. *Construction Management & Economics, 28*(6), 615–628.

Bresnen, M. (2010b). Keeping it real? Constituting partnering through boundary objects. *Construction Management and Economics, 28*(6), 615–628.

Bresnen, M., & Marshall, N. (2000). Building partnerships: Case studies of client-contractor collaboration in the UK construction industry. *Construction Management and Economics, 18,* 819–832.

Bresnen, M., & Marshall, N. (2002). The engineering or evolution of co-operation? A tale of two partnering projects. *International Journal of Project Management, 20*(7), 497–505.

Broom, J. (2002). *Procurement routes for partnering: A practical guide*. London: Thomas Telford.

Brown, M., Falk, A., & Fehr, E. (2004). Relational contracts and the nature of market interactions. *Econometrica, 72*(3), 747–780.

Brown, A. D., & Starkey, K. (1994). The effect of organizational culture on communication and information. *Journal of Management Studies (Oxford), 31*(6), 807–828.

Bygballe, L. E., Dewulf, G., & Levitt, R. E. (2015). The interplay between formal and informal contracting in integrated project delivery. *Engineering Project Organization Journal, 5*(1), 22–35.

Cannon, S., (2006, October). Procurement Management Journal, UK.

Chan, A. P., Chan, D. W., & Yeung, J. F. (2010). *Relational contracting for construction excellence: Principles, practices and case studies*. Abingdon: Spon Press.

Chan, E. H. W., & Tse, R. Y. C. (2003). Cultural considerations in international construction contracts. *Journal of Construction Engineering and Management, 129*(4), 375–381.

Chen, C. J. (2003). The effects of environment and partner characteristics on the choice of alliance forms. *International Journal of Project Management, 21*(2), 115–124.

Chen, L., & Manley, K. (2014). Validation of an instrument to measure governance and performance on collaborative infrastructure projects. *Journal of Construction Engineering and Management, 140*(5), 04014006. https://doi.org/10.1061/(ASCE)CO.1943-7862.0000834.

Chen, L., Manley, K., Lewis, J., Helfer, F., & Widen, K. (2018). Procurement and governance choices for collaborative infrastructure projects. *Journal of Construction and Engineering Management, 144*(8), 04018071.

Chen, I. J., & Popovich, K. (2003). Understanding customer relationship management (CRM) people, process and technology. *Business Process Management Journal, 9*(5), 672–688.

Chen, H., & Tien, H. C. (2007). Application of peer-to-peer network for real-time online collaborative computer aided design. *Journal of Computing in Civil Engineering, 21*(2), 112–121.

Cheng, E. W. L., Li, H., & Love, P. E. D. (2000). Establishment of critical success factors for construction partnering. *Journal of Management in Engineering, 16*(2), 84–92.

Cheung, S. O., Ng, T. S. T., Wong, S. P., & Suen, H. C. H. (2003). Behavioral aspects in construction partnering. *International Journal of Project Management, 21*(5), 333–343.

Chitkara, K. K. (2005). *Project Management—Planning, Scheduling and Controlling*. New Delhi: Tata McGraw Hill.

Cicmil, S., & Marshall, D. (2005). Insights into collaboration at the project level: Complexity, social interaction and procurement mechanisms. *Building Research and Information, 33*(6), 523–535.

Colledge, B. (2005). Relational contracting creating—value beyond the project. *Lean Construction Journal, 2*(1), 30–45.

Cravens, D., Cravens, K., & Piercy, N. (2002). Assessing the performance of strategic alliances: Matching metrics to strategies. *European Management Journal, 18*, 529–541.

Das, T. K., & Teng, B. S. (2001). Trust, control, and risk in strategic alliances: An integrated framework. *Organizations Studies, 22*(2), 251–283.

Davis, P. R., & Walker, D. H. T. (2009). Building capability in construction projects: A relationship-based approach. *Engineering, Construction and Architectural Management, 16*(5), 475–489.

Dawood, N., Akinsola, A., & Hobbs, B. (2002). Development of automated communication of system for managing site information using internet technology. *Automation in Construction, 11*(3), 552–572.

Demirkan, H. (2005). Generating design activities through sketches in multi-agent systems. *Automation in Construction, 14*(6), 699–706.

Dewulf, G., & Kadefors, A. (2012). Collaboration in public construction-contractual incentives, partnering schemes and trust. *Engineering Project Organization Journal, 2*(4), 240–250.

Diallo, A., & Thuillier, D. (2005). The success of international development projects: Trust and communication: An African perspective. *International Journal of Project Management, 23*(3), 237–252.

Doloi, H. (2009). Relational partnerships: The importance of communication, trust and confidence and joint risk management in achieving project success. *Construction Management and Economics, 27*(11), 1099–1109.

Donato, M. (2016). *The influence of resource dependency on collaboration in the construction supply chain*. Victoria: Victoria University.

Donato, M., Ahsan, K., & Shee, H. (2015). Resource dependency and collaboration in construction supply chain: Literature review and development of a conceptual framework. *International Journal of Project Management, 8*(3), 344–364.

Dozzi, P., Hartman, F., Tidsbury, N., & Ashraf, R. (1996). More stable owner–contractor relationships. *Journal of Construction Engineering and Management, ASCE, 122*(1), 30–35.

Drexler, J. A., & Larson, E. W. (2000). Partneering: Why project owner-contractor relationships change. *Journal of Construction Engineering and Management, ASCE, 126*(4), 293–297.

Duan, G., & Zhou, L. (2006). Study on product collaborative design system based on TeamCenter. *WSEAS Transactions on Computers, 5*(6), 1383–1388.

Egbu, C. O., & Robinson, H. S. (2005). Construction as a knowledge-based industry. In C. Anumba, C. Egbu, & P. Carillo (Eds.), *Knowledge Management construction* (pp. 31–49). Oxford: Blackwell Publishing.

Elmuty, D., & Kathawala, Y. (2001). An overview of strategic alliances. *Management Decision, 39*(3), 205–218.

Eriksson, P. (2010). Partnering: What is it, when should it be used, and how should it be implemented? *Construction Management and Economics, 28*(9), 905–917.

Eriksson, P. E. (2017). Procurement strategies for enhancing exploration and exploitation in construction projects. *Journal of Financial Management of Property and Construction, 22*(2), 211–230.

Ey, W., Zuo, J., & Han, S. (2014). Barriers and challenges of collaborative procurements: An exploratory study. *International Journal of Construction Management, 14*(3), 148–155.

Fellow, R. F., & Hancock, R. (1994). *Conflict resulting from cultural differentiation: An investigation of the new engineering contract*. Council of International Construction Research and Documentation Proc., Construction Conflict: Management and Resolution, CIB, Rotterdam, The Netherlands (pp. 259–267).

Fong, P. S. W., & Lung, B. W. C. (2007). Interorganizational teamwork in the construction industry. *Journal of Construction Engineering and Management, 133*(2), 157–168.

Gordon, C. M. (1994). Choosing appropriate construction contracting method. *Journal of Construction Engineering and Management, 120*(1), 196–210.

Gottlieb, S. C., & Jensen, J. S. (2012). Making sense of partnering: Discourses, governance and institutional change. *Engineering Project Organization Journal, 2*(3), 159–170.

Green, S. D. (1999) Partnering: the propaganda of corporatism? In S. O. Ogunlana (Ed.), *Profitable Partnering in Construction Procurement, CIB W92 and CIB TG23 Joint Symposium* (pp. 3–14). London: E&FN Spon.

Green, S. D., Fernie, S., & Weller, S. (2005). Making sense of supply chain management: A comparative study of aerospace and construction. *Construction Management and Economics, 23*(6), 579–593.

Greenham, P. (2007). *Alliancing: A Glimpse of the real world view*. Melbourne: Minter Ellison.

Hamza, A., Djebarni, R., & Hibberd, P. (1999). The implications of partnership success within the UK construction industry supply chain. In S. O. Ogunlana (Ed.), *Profitable Partnering in Construction Procurement, CIB W92 and CIB TG23 Joint Symposium* (pp. 39–46). London: E&FN Spon.

Harmon, K. M. J. (2003). Conflicts between owner and contractors: Proposed intervention process. *Journal of Management and Engineering, 19*(3), 121–125.

Hartmann, A., & Bresnen, M. (2011). The emergence of partnering in construction practice: An activity theory perspective. *Engineering project organization journal, 1*(1), 41–52.

Hartmann, A., Davies, A., & Frederiksen, L. (2010). Learning to deliver service-enhanced public infrastructure: Balancing contractual and relational capabilities. *Construction Management and Economics, 28*(11), 1165–1175.

Hauck, A. J., Walker, D. H. T., Hampson, K. D., & Peters, R. J. (2004). Project alliancing at national museum of Australia—Collaborative process. *Journal of Construction Engineering and Management, 130*(1), 143–152.

Hellard, R. (2002). *Project partnering: Principle and practice*. London: Thomas Telford.

Henisz, W. J., & Zelner, B. A. (2005). Legitimacy, interest group pressures, and change in emergent institutions: The case of foreign investors and host country governments. *Academy of Management Review, 30*, 361–382.
Henisz, W. J., Levitt, R. E., & Scott, W. R. (2012). Toward a unified theory of project governance: Economic, sociological and psychological supports for relational contracting. *Engineering Project Organization Journal, 2*(1–2), 37–55.
Hoetker, G., & Mellewigt, T. (2009). Choice and performance of governance mechanisms: Matching alliance governance to asset type. *Strategic Management Journal, 30*(10), 1025–1044.
Holzer, D., Tengono, Y., & downing, S. (2007). Developing a framework for linking intelligence from multiple professions in the AEC industry. In A. Dong, A. V. Moere, J. S. Gero (Eds.), *Computer aided architectural design futures (CAADFutures)*. Dordrecht: Springer.
Hughes, W., Champion, R., & Murdoch, J. (2015). *Construction contracts: Law and management*. New York: Routledge.
Ibrahim, C. K. I. C., Sabri, N. A. M., Belayutham, S., & Mahamadu, A. (2019). Exploring behavioural factors for information sharing in BIM projects in the Malaysian construction industry. *Built Environment Project and Asset Management, 9*(1), 15–28.
Ingirige, B., & Sexton, M. (2006). Alliances in construction: Investigating initiatives and barriers for long-term collaboration. *Engineering, Construction and Architectural Management, 13*(5), 521–535.
Jacobsson, M., & Roth, P. (2014). Towards a shift in mindset: Partnering projects as engagement platforms. *Construction Management and Economics, 32*(5), 419–432.
Jeffries, F. L., & Reed, R. (2000). Trust and adaptation in relational contracting. *Academy of Management Review, 25*(4), 873–882.
Judge, W. Q., & Dooley, R. (2006). Strategic alliance outcomes: A transaction cost perspective. *British Journal of Management, 17*, 23–37.
Kaplan, S., Schenkel, A., Krogh, G. V. & Weber, C. (2001). *Knowledge-based theories of the firm in strategic management: a review and extension*. MIT Sloan Working Paper, 4216-01.
Kent, D. C., & Becerik-Gerber, B. (2010). Understanding construction industry experience and attitudes toward integrated project delivery. *Journal of Construction Engineering and Management, 136*(8), 815–825.
Kling, E., & Burley, S. (1991). Joint ventures in Europe. In *European community joint ventures* (Chap. 1, p. 1). Butterworths.
Knott, T. (1996). *No business as usual: An extraordinary north sea result*. London: The British Petroleum Company.
Krishnan, R., Geyskens, I., & Steenkamp, J. (2016). The effectiveness of contractual and trust-based governance in strategic alliances under behavioral and environmental uncertainty. *Strategic Management Journal, 37*(12), 2521–2542.
Kumaraswamy, M. M., & Dissanayaka, S. M. (2001). Developing a decision support system for building project procurement. *Building and Environment, 36*(3), 337–349.
Kwan, A. Y., & Ofori, G. (2001). Chinese culture and successful implementation of partnering in Singapore's construction industry. *Construction Management and Economics, 19*(6), 619–632.
Kwok, H. C. A., Then, D., & Skitmore, M. (2000). Risk management in Singapore construction joint ventures. *Journal of Construction Research, 1*(2), 139–149.
Kwon, I. W., & Suh, T. (2005). Trust, commitment and relationships in supply chain management: A path analysis. *Supply Chain Management: An International Journal, 10*(1), 26–33.
Laan, A., Noorderhaven, N., Voordijk, H., & Dewulf, G. (2011). Building trust in construction partnering projects: An exploratory case-study. *Journal of Purchasing & Supply Management, 17*(2), 98–108.
Lahdenperä, P. (2012). Making sense of the multiparty contractual arrangements of project partnering, project alliancing and integrated project delivery. *Construction Management and Economics, 30*(1), 57–79.
Lahdenperä, P. (2017). Towards a coherent theory of project alliancing: Discovering the system's complex mechanisms yielding value for money. *Construction Economics and Building, 17*(2), 41–61.

Lamont, Z. (2001). Performing through partnering. *Civil Engineering, 144*(February), 3.
Larson, E., & Drexler, J. A. (1997). Barriers to project partnering: Report from the firing line. *Project Management Journal, 28*(2), 46–52.
Lee, J., Han, J., Paik, S., Kim, W., Jeon, H., & Choi, K. (2012). A method for the application of IPD to domestic construction industry through SWOT analysis. *Journal of the Architectural Institute of Korea, 281*(3), 99–106.
Leung, M. Y., Ng, S. T., & Cheung, S. O. (2004). Measuring construction project participant satisfaction. *Construction Management and Economics, 22*(3), 319–331.
Li, H., Cheng, E. W. L., & Love, P. E. D. (2000). Partnering research in construction. *Engineering, Construction and Architectural Management, 7*(1), 76–99.
Li, J., Lam, K., & Qian, G. M. (2001). Does culture affect behaviour and performance of firms? The case of joint ventures in China. *Journal of International Business Studies, 32*(1), 115–131.
Li, J., Zhou, C., & Zajac, E. J. (2009). Control, collaboration, and productivity in international joint ventures: Theory and evidence. *Strategic Management Journal, 30*(8), 865–884.
Lichtig, W. A. (2005). Sutter health: Developing a contracting model to support lean project delivery. *Lean Construction Journal, 2*(1), 105–112.
Lichtig, W. A. (2006). The integrated agreement for lean project delivery. *Constructions Lawyer, 26*(3), 1–8.
Ling, F., Peng Chong, T., Yan, N., Albert, T., & Asanga, G. (2015). Effect of adoption of relational contracting practices on relationship quality in public projects in Singapore. *Engineering, Construction and Architectural Management, 22*(2), 169–189.
Lloyd-Walker, B. M., Mills, A. J., & Walker, D. H. T. (2014). Enabling construction innovation: the role of a no-blame culture as a collaboration behavioural driver in project alliances. *Construction Management and Economics, 32*(3), 229–245.
Loosemore, M. (1999). International construction management research cultural sensitivity in methodological design. *Construction Management and Economics, 17*(5), 553–561.
Lou, M. M., Liew, P. S., Gu, N., & Ding, L. (2005). An agent approach to supporting collaborative design in 3D virtual worlds. *Automation in Construction, 14*(2), 189–195.
Love, P. E. D., Davis, P., Baccarini, D., Wilson, G., & Lopez, R. (2008). *Procurement selection in the public sector: A tale of two states.* Paper presented at 2008 CRC Construction Innovation Conference: Clients Driving Innovation. Gold Coast, Australia.
Love, E. D., & Gunasekarn, A. (1999). Learning Alliances: A customer – supplier focus for continuous improvement in manufacturing. *Industrial and Commercial Trading, 31*(3), 88–96.
Love, P. E. D., Mistry, D., & Davis, P. R. (2010). Price competitive alliance projects: Identification of success factors for public clients. *Journal of Construction and Engineering Management, 136*(9), 947–956.
Love, P. E. D., Skitmore, M., & Earl, G. (1998). Selecting a suitable procurement method for a building project. *Construction Management and Economics, 16*(2), 221–233.
Lu, P., Guo, S. P., Qian, L. M., He, P., & Xu, X. Y. (2015). The effectiveness of contractual and relational governances in construction projects in China. *International Journal of Project Management, 33*(1), 212–222.
Luu, D. T., Ng, S. T., & Chen, S. E. (2005). Formulating procurement selection criteria through case-based reasoning approach. *ASCE Journal of Computing in Civil Engineering, 19*(3), 269–276.
Manley, K. (2002). Partnering and alliancing on road projects in Australia and internationally. *Road and Transport Research, 11*(3), 46–60.
Matthews, O., & Howell, G. A. (2005). An integrated project delivery: An example of relational contracting. *Lean Construction Journal, 2*(1), 46–61.
Mead, S. T. (1999). *Communication effectiveness in intranet based construction projects.* Unpublished Ph.D. thesis, Department of Civil and building engineering, Loughorough, University, Loughorough, UK.
Middlebrooks, B. (2008). Integrated project delivery in practice. *Journal of Structural Engineering, 9*(12), 28–30.

Mockler, K. R. (2001). Making decisions on enterprise-wide strategic alignment in multinational alliances. *Management Decision, 39*(2), 90–98.
Molenaar, K. R., Sobin, N., & Antillon, E. I. (2010). A synthesis of best value procurement practices for sustainable design-build projects in the public sector. *Journal of Green Building, 5*(4), 148–157.
Morledge, R., Smith, A., & Kashiawagi, D. T. (2006). *Building procurement*. Oxford: Blackwell Publishing Ltd.
Morwood, R., Scott, D., & Pitcher, I. (2008). *Alliancing: A participant's guide*. Brisbane: Maunsell AECOM.
Naoum, S., & Mustapha, F. (1995). Relationship between the building team, procurement methods and project performance. *Journal of Construction Procurement, 1*(1), 50–63.
Ng, S. T., Rose, T. M., Mak, M., & Chen, S. E. (2002). Problematic issues associated with project partnering – The contractor perspective. *International Journal of Project Management, 20*, 437–449.
Ngowi, A. B., & Pienaar, E. (2005). Trust factor in construction alliances. *Building Research & Information, 33*(3), 267–278.
Ning, Y., & Ling, F. Y. Y. (2015). The effects of project characteristics on adopting relational transaction strategies. *International Journal of Project Management, 33*(5), 998–1007.
Norwood, S. R., & Mansfield, N. R. (1999). Joint venture issues concerning European and Asian construction markets of the 1990's. *International Journal of Project Management, 17*(2), 89–93.
Osipova, E., & Eriksson, P. E. (2011). How procurement options influence risk management in construction projects. *Construction Management and Economics., 29*(11), 1149–1158.
Otter, A. F. H. J. (2005). *Design team communication using a project website*. Ph.D. thesis, Bouwstenen 98, Technische Universiteit Eindhoven, Eindhoven.
Parkhe, A. (1993). Strategic alliance structuring: a game theoretic and transaction cost examination of interfirm cooperation. *Academy of Management Review, 36*(4), 794–829.
Parrish, K., Wong, J. M., Tommelein, I. D., & Stojadinovic, B. (2008). Set-based design: A case study on innovative hospital design. In P. Tzortzopoulos & M. Kagioglou (Eds.), *Proceedings of the 16th Annual Conference of the International Group for Lean Construction (IGLC-16)* (pp. 413–423). International Group of Lean Construction, Manchester.
Patel, T. (2007). The role of dynamic cultural theories in explaining the viability of international strategic alliances: A focus on Indo-French alliances. *Management Review, 45*(10), 1532–1559.
Perlberg, B. (2009). Contracting for integrated project delivery: ConsensusDOCS. In V. O. Schinnerer, Md. Chevy Chase (Eds.), *Proceedings of the 48th Annual Meeting of Invited Attorneys*.
Perrenoud, A., Lines, B. C., Savicky, J., & Sullivan, K. T. (2017). Using best-value procurement to measure the impact of initial risk-management capability on qualitative construction performance. *Journal of Management and Engineering, 33*(5), 1–8.
Perry, M., & Sanderson, D. (1998). Coordinating joint design work: The role of communication and artefacts. *Design Studies, 19*(3), 273–288.
Pett, T. L., & Dibrell, C. C. (2001). A process model of global strategic alliance formation. *Business Process Management Journal, 7*(4), 349–361.
Phua, F. T. T., & Rowlinson, S. (2003). Cultural differences as an explanatory variable for adversarial attitudes in the construction industry: The case of Hong Kong. *Construction Management and Economics, 21*(7), 777–785.
Poppo, L., & Zenger, T. (2002). Do formal contracts and relational governance function as substitutes or complements. *Strategic Management Journal, 23*(8), 707–725.
Pryke, S. D. (2004). Analysing construction project coalitions: Exploring the application of social network analysis. *Construction Management and Economics, 22*(8), 787–797.
Pryke, S. D. (2005). Towards a social network theory of project governance. *Construction Management and Economics, 23*(9), 927–939.
Pryke, S., & Pearson, S. (2006). Project governance: Case studies on financial incentives. *Building Research and Information, 34*(6), 534–545.

Rahman, M., & Kumaraswamy, M. M. (2002). Joint risk management through transactionally efficient relational contracting. *Construction Management and Economics, 20*(1), 45–51.

Rahman, M. M., & Kumaraswamy, M. M. (2004). Contracting relationship trends and transitions. *Journal of Management in Engineering, 20*(4), 147–161.

Rahman, M. M., & Kumaraswamy, M. M. (2005). Relational selection for collaborative working arrangements. *ASCE Journal of Construction Engineering and Management, 131*(10), 1087–1098.

Rahman, M. M., & Kumaraswamy, M. M. (2012). Multi-country perspectives of relational contracting and integrated project teams. *Journal of Construction Engineering and Management, 138*(4), 469–480.

Rahman, M. M., Kumaraswamy, M. M., & Ling, F. Y. Y. (2007). Building a relational contracting culture and integrated teams. *Canadian Journal of Civil Engineering, 34*(1), 75–88.

Ranganathan, C., & Lerpittayapoom, N. (2002). Towards a conceptual framework for understanding strategic alliance in e-commerce. In *Proceedings of the 35th Hawaii International Conference on System Sciences*, Hawaii, January 7–10.

Rao, B. P., & Swaminathan, V. (1995). Uneasy alliances: Cultural incompatibility or culture shock. In *Proceedings of the Association of Management 13th Annual International Conference*, Vancouver, Canada.

Ravasi, D., & Schultz, M. (2006). Responding to organizational identity threats: Exploring the role of organizational culture. *Academy Management of Journal, 49*(3), 433–458.

Ren, Z., Anumba, C. J., Hassan, T. M., Augenbroe, G., & Mangini, M. (2006). Collaborative project planning: A novel approach through an e-engineering hub—A case study of seismic risk analysis. *Computers in Industry, 57*(3), 218–230.

Reuer, J., & Devarakonda, S. (2016). Mechanisms of hybrid governance: Administrative committees in non-equity alliances. *Academy Management of Journal, 59*(2), 510–533.

Robson, M. (2002). Partner selection in successful strategic alliances: The role of cooperation. *Journal of General Management, 28*(1), 1–15.

Ross, J. (2006). *Project alliancing: Practitioners' guide*. Melbourne: The Department of Treasury and Finance.

Rowlinson, S., Cheung, F., Simons, R., & Rafferty, A. (2006). Alliancing in Australia—no-litigation contracts: a tautology? *Journal of Professional Issues in Engineering Education and Practice, 132*(1), 77–81.

Sakal, M. (2005). Project alliancing: A relational contracting mechanism for dynamic projects. *Lean Construction Journal, 2*(1), 67–79.

Salvato, C., Reuer, J., & Battigalli, P. (2017). Cooperation across disciplines: A multilevel perspective on cooperative behaviour in governing interfirm relations. *The Academy of Management annals, 11*(2), 960–1004.

Sha, K. (2016). Relational contracting in China's building sector: Potentialities and challenges. *International Journal of Architecture, Engineering and Construction, 5*(4), 207–216.

Sillars, D. N., & Kangari, R. (2004). Predicting organizational success within a project-based joint venture alliance. *Journal of Construction Engineering and Management, 130*(4), 500–508.

Sive, T. (2009). *Integrated project delivery: Reality and promise, a strategist's guide to understanding and marketing IPD*. Society for Marketing Professional Services Foundation.

Sørensen, J. B. (2002). The strength of corporate culture and the reliability of firm performance. *Administrative Science Quarterly, 47*(1), 70–91.

Stock, G. N., McFadden, K. L., & Gowen, C. R. (2007). Organizational culture, critical success factors, and the reduction of hospital errors. *International Journal of Production Economics, 106*(2), 368–392.

Suprapto, M., Bakker, H. L., & Mooi, H. G. (2015). Relational factors in owner-contractor collaboration: The mediating role of teamworking. *International Journal of Project Management, 33*(6), 1347–1363.

Taylor, J. B. (1999). A historical analysis of monetary policy rules. In J. B. Taylor (Ed.), *Monetary Policy Rules*, Chicago: University of Chicago Press.

Tennant, S., & Fernie, S. (2012). The commercial currency of construction framework agreements. *Building Research & Information, 40*(2), 209–220.
Thomas, S. R., Tucker, R., & Kelly, W. (1999). Compass: an assessment tool for improving project team communications. *Project Management Journal, 30*(4), 15–23.
Turner, J. R., & Müller, R. (2003). On the nature of the project as a temporary organization. *International Journal of Project Management, 21*(1), 1–8.
Tyler, A., & Matthews, J. (1996)/ *An evaluation of definitions and key elements of partnering.* Paper presented at the RICS International Research Conference Cobra 1996, University of West England, Bristol, 19–20 September.
Vaaland, T. I. (2004). Improving project collaboration: Start with the conflicts. *International Journal Project Management, 22*(6), 447–454.
Van Vijfeijken, H., Kleingeld, A., Van Tujil, H., Algera, J. A., & Thierry, H. (2002). Task complexity and task, goal and reward independence in group performance management: A prescriptive model. *European Journal of Work and Organizational Psychology, 11*(3), 363–383.
Walker, D., & Hampson, K. (eds). (2003). *Procurement strategies: A relationship-based approach.* Oxford: Blackwell Science.
Walker, D. H. T., & Rahamani, F. (2016). Delivering a water treatment plant project using a collaborative project procurement approach. *Construction Innovation. Information, Process, Management, 16*(2), 158–184.
Wheelen, T. L., & Hungar, D. J. (2000). *Strategic management and business policy* (pp. 125–134). New York, NY: Addison Wesley.
Wikforss, Ö., & Lofgren, A. (2007). Rethinking communication in construction. *Journal of Information Technology in Construction, 12,* 337–346.
Winch, G. M. (2010). *Managing construction projects.* Chichester, West Sussex: Wiley.
Wong, P. S. P., Cheung, S. O., & Ho, P. K. M. (2005). Contractor as trust initiator in construction partnering—Prisoner's dilemma perspective. *Journal of Construction Engineering and Management, 131*(10), 1045–1053.
Wood, G. (2005). *Partnering practice in the relationship between clients and main contractors.* RICS Paper series, Vol. 5 No. 2, RICS, London.
Wood, G., & McDermott, P. (1999). Searching for trust in the UK construction industry: an interim view. In S. O. Ogunlana (Ed.), *Profitable Partnering in Construction Procurement, CIB W92 and CIB TG23 Joint Symposium,* London, E&FN Spon (pp. 107–16).
Xie, X., Thorpe, A., & Baldwin, A. N. (2000). A survey of communication issues in construction design. In Akintoye, A. (Ed.), *16th Annual ARCOM Conference,* 6–8 September 2000, Glasgow Caledonian University. Association of Researchers in Construction Management (Vol. 2, pp. 771–780).
Xie, C., Wu, D., Luo, J., & Hu, X. (2010). A case study of multi-team communications in construction design under supply chain partnering. *Supply Chain Management: An International Journal, 15*(5), 363–370.
Xue, X., Shen, Q., & Ren, Z. (2010). Critical review of collaborative working in construction projects: Business environment and human behaviors. *Journal of Management in Engineering, 26*(4), 196–208.
Yasuda, H. (2005). Formation of strategic alliances in high-technology industries: A comparative study of the resource-based theory and the transaction-cost theory. *Technovation, 25,* 763–770.
Yeung, K., Zhou, H., Yeung, A. C. L., & Cheng, T. C. E. (2012). The impact of third-party logistics providers' capabilities on exporters' performance. *International Journal of Production Economics, 135*(2), 741–753.
Young-Ybarra, C., & Wiersema, M. (1999). Strategic flexibility in information technology alliances: The influence of transaction cost economics and social exchange theory. *Organization Science, 10*(4), 439–459.
Zimina, D., Ballard, G., & Pasquire, C. (2012). Target value design: Using collaboration and a lean approach to reduce construction cost. *Construction Management and Economics, 30*(5), 383–398.

Chapter 6
Nature of Barriers in Construction Communication Performance

Abstract Effective communication has become an essential and indispensable factor to construction project delivery across various procurement typologies. In spite of this understanding, poor communication continues to be a persistent problem in project delivery. Construction project delivery is said to be an information-intensive endeavour and as such the urge of exploring ways of improving communication effectiveness is an urgent necessity. Communication performance assessment is considered as a veritable approach to identifying communication issues in construction as a prelude to developing strategies for improvement. Evidence from literature points to the fact that communication performance lacks a consistent understanding across various domains where it has been used, conceptualized as well as measured. This chapter provides an understanding of communication performance in construction and how it is measured. An insight into the barriers to communication is also provided. This offers useful knowledge that can impact on nature of communication strategies and process that can influence communication performance in project delivery.

Keywords Effective communication · Communication performance · Communication barriers · Procurement typologies

6.1 Introduction

Construction process is said to be information intensive making effective communication performance central to process efficiency and success. However, numerous studies have reported that communication in construction suffers from communication problems and barriers with varying perception about these barriers. In the global construction industry today, organizations and businesses are forming diverse strategic alliances, partnering, coalition and joint ventures, making communication moving beyond individual level and becoming more complex riddled with numerous barriers which must be understood if efforts can engender improvement in communication effectiveness. General literature classifies construction communication among the domain of organizational communication and further acknowledges that there are several barriers to effective communication in construction environment. However, these barriers are theoretically conceptualized in varied forms. The idea behind the

T. E. Kwofie et al., *Effective Construction Project Delivery*,
https://doi.org/10.1007/978-3-030-49374-5_6

gaining an understanding of the barriers to communication in non-traditional procurement is for it to serve as a variable approach towards enhancing, improving and increasing the predictability of communication performance in project delivery. This is with the view of eliminating the barriers to communication in project teams.

6.2 Understanding the Concept of Communication Process and Communication Performance

Communication process has been conceptualized as being linear and nonlinear in several scholarly presentations. Effectiveness and efficiency in the communication process and networks for among project teams is becoming increasingly important due to the growing technical and organizational complexity of construction projects. It is also considered as a panacea for improving collaborative working but this is proving to be elusive in practice in the industry especially across all procurement typologies (Damodaran and Shelbourn 2006; Emmitt and Gorse 2007; Gorse and Emmitt 2003; Gorse 2002).

Shannon and Weaver (1949) developed a model to explain human communication which is seen as the earliest attempt on the subject. Here, communication was seen as linear interactions where a sender encodes a message and sent to a receiver who decodes it through a medium in one direction. In one-direction mean, the model excludes feedback which is an important component of communication and human interaction. Schermerhorn et al. (1994) expanded this model to include a feedback loop between the sender and the receiver thus giving communication a two-direction approach. Though these communication models were extensively pursued in psychology, linguisyics and sociology, they became the building blocks for developing construction communication (see Emmitt and Gorse 2007; Dainty et al. 2006; Thomas et al. 1998; Emmitt and Gorse 2003, Pietroforte 1997; Senaratne and Ruwanpura 2016).

The communication process in construction has been repeatedly been described and conceptualized as being dynamic and complex composing of intrapersonal, interpersonal and interorganizational interactions influenced by social and task-based interactions (Gorse and Emmitt 2007; McLeod and Kettner-Polley 2004). Construction communication process apart from being dynamic has had communicators, information and message, network and media as the key elements in the process (see Emmitt and Gorse 2003; Gorse 2002; Otter and Emmitt 2007; Thomas et al. 1998). Xie et al. (2010) and Lou et al. (2005) indicated that construction communication process varies across project life cycle phases as participants get off and join the team from one phase to the other and also across various procurement typologies.

It is worth noting that, in the construction industry today, communication process has extended to include a virtual and digital dimension where information has included acoustic, visual or digital messages among project teams (Dainty et al. 2006; Liu 2009; Xie et al. 2010; Kwofie et al. 2016). Communication process in construction predominantly takes a two-way process where the information encoded must be sent through a medium to the receiver and decoded for the sender to receive a feedback (Dainty et al. 2006; Thomas et al. 1998). By conceptualizing construction communication process this way, it integrates the physical and social context of the working environment (context) of people or groups (Dainty et al. 2006; Xie et al. 2010). Given that, construction communication follows the precepts of human interactions which mean that, as typical of human interactions, feedbacks aid in affirming the information in the interactions has been satisfactorily conveyed.

When communication process is perceived in two ways, it presents a more dynamic or iterative concept allowing for continuous receipt of feedback that is considered crucial towards communication effectiveness (Dainty et al. 2006; Gorse and Emmitt 2007).

6.3 Environment of Construction Project Communication

Communication is regarded as an important instrument that informs, regulates, integrates, manages, persuades and socializes (Richmond and McCroskey 1992). However, it is a common knowledge that the ability of communication to perform these functions is dependent on the context or the environment of communication. Tracing the related literature on communication environment or context in construction is significant providing some insights into the underlying elements of communication barriers experienced in communication networks, media and the process among the project team.

Dainty et al. (2006) affirmed that construction project environment exhibits enormous diversity of agents involved with different specialities, circumstances and ways of understanding situations which often creates a conflict and adversarial situations. Mondrup et al. (2012) and Morreale et al. (2006) hinted that construction project participants form teams or groups with increasingly complex communication networks which have implications for communication management and information sharing. Xue et al. (2010) accounted that the construction project delivery is showing changing business environment characterized by tense competitiveness and wide global links. This evolution is in addition to it already known fragmentation and non-collaborative nature (Dawood et al. 2002; Manley and Chen 2016; Kwofie et al. 2014; Xie et al. 2010).

Project participants have significant roles to play in a complex network of tasks and social interactions; hence, understanding the project environment as an interconnected network of actors is most appropriate (Gorse and Emmitt 2007; Otter and Emmitt 2007). Given this nature of the project environment in construction, it is right to assert that ensuring effective process and team effectiveness requires establishing effective and efficient interorganization collaborative and integrated management systems which is highly dependent on effective communication and information sharing. This is against the background of emerging formation of strategic alliance and partnering in construction supply chain which is from widely distributed geographic areas compared to the conventional co-located supply chains (Xue et al. 2010; Xie et al. 2010).

Construction project environment also exhibits intense cultural diversity which impacts on team dynamics and procurement systems (Rowlinson et al. 2008). It is worth to say that increasing internationalization of many construction industry has sought to exacerbate the cultural influence. This obviously influences how team interacts, shares information, builds trust, mutual respect and conflict resolution strategy (Xue et al. 2007; Chan et al. 2004; Hauck et al. 2004). Construction project environment is seen as being peculiar in it intense reliance on information and communication (Cui et al. 2018; Kwofie et al. 2016; Dainty et al. 2006). The project delivery process also often experiences a series of groups, teams, organizations and networks that are temporary in nature generated by relationships and interactions that are constantly changing reflecting the dynamism of the work environment (Goczol and Scoubeau 2003; Pietroforte 1997).

Ceric (2014) also alluded that the construction industry covers an array of projects with each project being unique in nature as it involves myriads of interrelated activities, tasks and work. The construction industry is revered for being a highly sector utilizing series of defined protocols that can facilitate communication (Goczol and Scoubeau 2003). Given the rapidly changing global business industry, it can be propounded that it is not in doubt that the construction industry needs to be adaptable to new environment necessary to maintain its competitiveness, core business and improve coordination and optimize performance (Cheng et al. 2001, 2013).

In spite of the generally accepted fact that improvement in the sector is urgent and imminent, the construction industry has focused on process and product improvements ahead of the need to improve the complex interorganizational and interpersonal relationships that define the industry's culture (see Emmitt and Gorse 2003; Gorse and Emmitt 2007). Project team or group members have different and varied background and often share different perspectives of situations. These characteristics of the construction project environment induce significant barriers to the process, behaviours, interactions collaborative working, team dynamics and effectiveness (Xue et al. 2005; Chan et al. 2004; Hauck et al. 2004; Cheng et al. 2001, 2013). An in-depth understanding of the inherent barriers into construction communication has extensively been skewed to traditional and conventional procurement systems which differ significantly from relational contracting that has become dominant option in the global industry.

6.4 Profile and Typologies of Barriers in Construction Communication

A synthesis of existing literature surrounding human organizational communication provide evidence in support of incidence of various typologies of barriers to effective communication which construction communication is no exception (see Pietroforte 1997; Kwofie et al. 2016; Dainty et al. 2006; Otter and Emmitt 2007; Xie et al. 2010;

Xie et al. 2000; Dennis and Kinney 1998). However, analysis and identification of these challenges are differently conceptualized and are disjointed in the construction project delivery. Notably, there is the evidence of describing barriers to communication as problems or challenges (Xie et al. 2010; Xie et al. 2000; Xie 2002; Kwofie et al. 2014). This conceptualization may be erroneous given that by following the Thomas et al. (1999) COMPASS tool for communication challenges assessment in construction which has extensively been used; it can be noted that barriers is just one of the problems of communication effectiveness. With the impact barriers make to communication effectiveness, it is sad to note that this has not been explored extensively. Expanding the scope of barriers to communication in construction beyond the current knowledge can be deemed as a good start and provides a platform for further research in the subject area. Guevara and Boyer (1981) highlighted barriers to effective communication among other communication problems in construction and intimated that the problems of barriers were prevalent in the construction industry. Kwofie et al. (2015, 2016) conceptualized communication problems inherent from the unique attributes of projects from information composition and information flow and found evidence of barriers in both the information flow and composition. These studies appear to be a pioneer one exploring communication problems form both information flow and composition jointly.

Thomas et al. (1998) defined barriers to communication in construction as any form of interpersonal, accessibility, logistic factors interfering with communication between participants in project delivery. Mead (1999) revealed that the use of COMPASS in evaluating communication problems has gained credence and been very effective, but the area of evaluating barriers should be broached with additional questions. Various studies have investigated various typologies of barriers to communication in construction. From these existing knowledge, it can be synthesized that behavioural, logistics, skills, organizational, cultural diversity, project characteristics project environment (context) and procurement systems have been typologies of barriers that have been explored (see Ochieng and Price 2009, 2010; Kwofie et al. 2015; Liu 2009; Xie et al. 2010; Thomas et al. 1998; Diallo and Thuiller 2005; Gorse and Sansderson 2007; Gorse and Emmitt 2007).

Gorse and Emmitt (2007) indicated that project team participants adopt various forms of communication behaviour in both task and relational interactions and information sharing. Their findings also showed that poor communication behaviour forms a greater barrier to task-based interactions. Hence, high levels of task-based interactions and low levels of socio-emotional interaction require the optimum communication behaviours to overcome notable barriers in the communication projects especially in meetings and asynchronous interactions (Gorse and Emmitt 2007).

Mead (1999) and Otter (2005) intimated that intranet, Internet and project website have become significant resources aiding project communication and information management. Generally, it can be affirmed that computing and communication technologies continue to play a major role towards improving design communication (Demirkan 2005). However, this has not been the case globally. In spite of the fact that, in todays global construction industry, the Internet and project website have been widely accepted tools in the direction of information management and communication, lack of access to reliable Internet have become a logistical barrier to timely access to project information. This is affirmed in a study by Kwofie et al. (2015) that, in Ghana and other developing countries, most construction sites are not linked to reliable Internet sources making access to shared project information a huge problem resulting in continuous delays in decisions and implementation.

The revelation provided by Kwofie et al. (2015b, 2016) indicates that in information composition for communication, senders and receivers must develop task functional skills and psycho-social skill to have the right competencies for effective communication. Hence, lack of these skills constitute a significant barrier to effective communication performance among project teams. This is because the more effective a communication is can be drawn from the competencies of the communicators (Bagarić and Djigunović 2007; Salleh 2008).

Organizational barriers to effective communication take the form of omitting someone from the communication chain or sending incomplete or over simplified information due to the organizational structure or operational process (Xue et al. 2010; Liu 2009). Various forms of organizational setup for project management and delivery exhibit unique attributes that significantly impact on communication process, strategy, procedure and protocol (Xue et al. 2010; Xie et al. 2010; Shohet and Frydman 2003; Murray et al. 2000). Against this, organizational setup with clear object, high-level trust and openness tend to facilitated communication and information sharing in project delivery than one with which is adversarial and high level of mistrust and lack of collaboration (Xie et al. 2010). From the work of Xie et al. (2010), an empirical assentuation was given which indicated that project organizational setups which embrace innovation and promote initiatives enhance effective communication hence eliminating all forms of organizational barriers.

Various research findings have given theoretical and empirical credence to the fact that the nature, characteristics, attributes and complexities of projects pose obvious barriers to information flow, interactions and communication mechanism (see Kwofie et al. 2015; Kwofie 2015; Luck 2007; Cheng et al. 2001; Wikforss and Alexander 2007). The apparent barrier inherent from the unique attributes of construction project is the inability of applying known and established communication techniques and models across all project typologies. Kwofie et al. (2015) provided empirical evidence to the unique nature of mass housing projects, and thus, the nature poses unique communication barriers entirely in variance to that of traditional one-off construction projects.

Love et al. (2001) initiated steps to explore the structural flexibility of construction project procurement and delivery process. The evidence gathered point to the fact that conventional procurement systems are more rigid and adversarial to modifications and adjustments as oppose to relational contracting (non-traditional systems) and this poses a notable barrier to team communication and information sharing. Xie et al. (2010) revealed that relational forms of contracting such as partnering and alliancing facilitated effective communication compared to traditional contracting arrangements.

The commentary provided by Ochieng and Price (2009) gives a theoretical assertion that the international and diversed nature of project teams make them multi-cultural and culturally diversed in nature which brings new challenges in their task-based and social-based interactions and communication. Against this background, a good demonstration of an awareness of cultural variation is said to be a precursor for eliminating cultural-related barriers to effective communication (Ochieng and Price 2010; Diallo and Thuillier 2005). The works of Robbins (2001) and Gorse (2002) also point to the fact that the multi-disciplinary nature of project teams coupled with participants belonging to different organizations exposes project team communication to different organizational cultures which consequently induce a form of barrier to their information sharing and communication. This organization cultural diversity often bring along the skill diversity, different levels of preferences that affect the rate of adoption of given communication tools (Robbins 2001; Gorse 2002).

6.5 Summary

Barrier to construction communication was previously conceptualized and diffused among the problems of communication overly presented in literature. From this conceptualization, it was clearly noted that there existed an ineptitude in the approach, inadequacy and lack of depth in the measure. In this chapter, an in-depth understanding of barriers to communication in construction has been provided and clarity given to the typologies of barriers experienced among construction project teams in both their task-based and social-based interactions and communication. From a synthesis of existing knowledge, behavioural, logistics, organizational, cultural diversity, project characteristics project environment (context) and procurement systems have been typologies of barriers in construction communication. However, understanding of nature and extent of these barriers must be pursued across various forms of procurement systems to advance the existing knowledge theoretically and practically.

References

Bagarić, V., & Djigunović, J. M. (2007). Defining communicative competence. *Metodika, 8*(1), 94–103.

Ceric, A. (2014). Minimizing communication risk in construction: A Delphi study of the key role of project managers. *Journal of Civil Engineering and Management, 20,* 829–838.

Chan, A. P. C., Chan, D. W. M., Chiang, Y. H., Tang, B. S., Chan, E. H. W., & Ho, K. S. K. (2004). Exploring critical success factors for partnering in construction projects. *Journal of Construction Engineering and Management, 130*(2), 188–198.

Cheng, J. H., Chen, S. W., & Chen, F. Y. (2013). Exploring how inter-organizational relational benefits affect information sharing in supply chains. *Information Technology and Management, 14*(4), 283–294.

Cheng, E. W. L., Li, H., Love, P. E. D., & Irani, Z. (2001). Network communication in the construction industry. *International Journal of Corporate Communications, 6,* 61–70.
Cui, C., Liu, Y., Hope, A., & Wang, J. (2018). Review of studies on the public–private partnerships (PPP) for infrastructure projects. *International Journal of Project Management, 36*(5), 773–794.
Dainty, A., Moore, D., & Murray, M. (2006). *Communication in construction-theory and practice.* Abingdon, Oxon: Taylor & Francis.
Damodaran, L., & Shelbourn, M. (2006). Collaborative working—The elusive vision. *Architectural Engineering and Design Management, 2*(4), 227–243.
Dawood, N., Akinsola, A., & Hobbs, B. (2002). Development of automated communication of system for managing site information using internet technology. *Automation in Construction, 11*(3), 552–572.
Demirkan, H. (2005). Generating design activities through sketches in multi-agent systems. *Automation in Construction, 14*(6), 699–706.
Dennis, A. R., & Kinney, S. T. (1998). Testing media richness theory in new media: The effects of cues, feedback, and task equivocality. *Information Systems Research, 9*(2), 160–175.
Diallo, A., & Thuillier, D. (2005). The success of international development projects: Trust and communication: An African perspective. *International Journal of Project Management, 23*(3), 237–252.
Emmitt, S., & Gorse, C. A. (2003). *Construction communication.* Oxford: Blackwell Publishing.
Emmitt, S., & Gorse, C. A. (2007). *Communication in construction teams.* London: Taylor & Francis.
Goczol, J., & Scoubeau, C. (2003). Corporate communication and strategy in the field of projects. *Corporate Communication an International Journal, 8*(1), 60–66.
Gorse, C.A. (2002). *Effective interpersonal communication and group interaction during construction management and design team meetings.* Ph.D. thesis, University of Leicester.
Gorse, C. A., & Emmitt, S. (2003). Investigating interpersonal communication during construction progress meetings: challenges and opportunities. *Engineering, Construction and Architectural Management, 10*(4), 234–244.
Gorse, C. A., & Emmitt, S. (2007). Communication behaviour during management and design team meetings: a comparison of group interaction. *Construction Management and Economics, 25*(11), 1197–1213.
Gorse, C. A., & Sansderson, A. M. (2007). Exploring group work dynamics. In D Boyd (Ed.), *Proceedings of 23rd Annual ARCOM Conference,* 3–5 September 2007, Belfast, UK, Association of Researchers in Construction Management (pp. 295–296).
Guevara, J. M., & Boyer, L. T. (1981). Communication problems within construction. *Journal of Construction Engineering, American Society of Civil Engineers, 107*(4), 552–557.
Hauck, A. J., Walker, D. H. T., Hampson, K. D., & Peters, R. J. (2004). Project alliancing at national museum of Australia—Collaborative process. *Journal of Construction Engineering and Management, 130*(1), 143–152.
Kwofie, E. (2015). *Contribution of unique features of mass housing projects to project team communication performance,* unpublished PhD thesis, Kwame Nkrumah University of Science and Technology, Department of Building Technology, Kumasi.
Kwofie, T. E., Fugar, F., & Adinyira, E. (2015). Contribution of multiple construction site management features to project team communication effectiveness: The case of mass housing projects. *Engineering Project Organization Journal, 5,* 180–193.
Kwofie, T. E., Fugar, F., & Adinyira, E. (2015b). Identifying the communication competency behaviors of mass housing project teams in developing countries: An exploratory study. *Covenant Journal of Research in the Built Environment (CJRBE), 3*(1), 22–43.
Kwofie, T. E., Fugar, F., & Adinyira, E. (2016). Modelling the effect of housing design unit contract packaging on mass housing project team communication performance. *Journal of Construction in Developing Countries, 21*(1), 35–50.
Kwofie, T. E., Fugar, F. D.K, Adinyira, E., & Ahadzie, D. K. (2014). A conceptual framework for evaluating communication performance among mass housing project team. In O. Ejohwomu &

O. Oshodi (Eds.), *Proceedings of 3rd International Conference on Infrastructure Development in Africa*, 17–19 March, 2014, Abeokuta, Nigeria (pp. 6–21).
Liu, Y. (2009). *Critical factors for managing project team communication at the construction stage*. Ph.D. Thesis. Submitted to the Polytechnic University of Hong Kong.
Lou, M. M., Liew, P. S., Gu, N., & Ding, L. (2005). An agent approach to supporting collaborative design in 3D virtual worlds. *Automation in Construction, 14*(2), 189–195.
Love, P. E. D., Irani, Z., Li, H., & Cheng, E. (2001). An empirical analysis of IT/IS evaluation in construction. *Journal of Construction Information Technology, 8*(1), 15–27.
Luck, R. (2007). Using artefacts to mediate understanding in design conversations. *Building Research & Information, 35*(1), 28–41.
Manley, K., & Chen, L. (2016). The impact of client characteristics on the time and cost performance of collaborative infrastructure projects. *Engineering, Construction and Architectural Management, 23*(4), 511–532.
McLeod, P. L., & Kettner-Polley, R. B. (2004). Contributions of psychodynamic theories to understanding small groups. *Small Group Research, 35*(3), 333–361.
Mead, S. T. (1999). *Communication effectiveness in intranet based construction projects*, Unpublished Ph.D thesis, Department of Civil and building engineering, Loughorough, University, Loughorough, UK.
Mondrup, T. F., Karlshøj, J., & Vestergaard, F. (2012). Communicate and collaborate by using building information modeling. In *International Council for Research and Innovation in Building and Construction (CIB) W078 Conference*, Beirut, 17–19 October.
Morreale, S. P., Spitzberg, B. H., & Barge, J. K. (2006). *Human communication: Motivation, knowledge, and skills* (2nd ed.). Belmont, CA: Thomson Learning.
Murray, M. D., Tookey, J. E., Langford, D. A., & Hardcastle, C. (2000). Project communication variables: A comparative study of US and UK construction industry perceptions. In *ARCOM Conference Proceedings*, Glasgow, UK (pp. 813–22).
Ochieng, E. G., & Price, A. D. (2009). Framework for managing multicultural project teams. *Engineering, Construction and Architectural Management, 16*(6), 527–543.
Ochieng, E. G., & Price, A. D. (2010). Managing cross-cultural communication in multi-cultural construction project teams: The case of Kenya and UK. *International Journal of Project Management, 28*(5), 449–460.
Otter, A. F. H. J. (2005). *Design team communication using a project website*, PhD thesis, Bouwstenen 98, Technische Universiteit Eindhoven, Eindhoven.
Otter, A., & Emmitt, S. (2007). Exploring effectiveness of team communication balancing synchronous and asynchronous communication in design teams. *Engineering, Construction and Architectural Management, 14*(5), 408–419.
Pietroforte, R. (1997). Communication and governance in the building process. *Construction Management and Economics, 15,* 71–82.
Richmond, V. P., & McCroskey, J. C. (1992). *Organisational communication for survival*. Englewood Cliffs, NJ: Prentice Hall.
Robbins, S. P. (2001). *Organizational behavior*. Englewood Cliffs, NJ: Prentice-Hall.
Rowlinson, S., Walker, D. H. T., & Cheung, F. Y. K. (2008). Culture and its impact upon project procurement. In D. H. T. Walker & S. Rowlinson (Eds.), *Procurement systems: A cross-industry project management perspective*. New York, NY: Taylor and Francis.
Salleh, L. M. (2008). Communication competence: A Malaysian perspective. *Pacific and Asian Communication Association, 11*(3), 303–312.
Schermerhorn, J. R., Jr., Hunt, J. G., & Osborn, R. N. (1994). *Managing organisational behaviour*. New York, NY: Wiley.
Senaratne, S., & Ruwanpura, M. (2016). Communication in construction: A management perspective through case studies in Sri Lanka. *Architectural Engineering and Design Management, 12*(1), 3–18.
Shannon, C. E., & Weaver, W. (1949). *The mathematical theory of communication*. Chicago: University of Illinois Press.

Shohet, I. M., & Frydman, S. (2003). Communication patterns in construction at construction manager level. *Journal of Construction Engineering and Management, 129*(5), 570–577.

Thomas, S. R., Tucker, R., & Kelly, W. (1998). Critical communications variables. *Journal of Construction Engineering Management, 129*(1), 58–66.

Thomas, S. R., Tucker, R., & Kelly, W. (1999). Compass: An assessment tool for improving project team communications. *Project Management Journal, 30*(4), 15–23.

Wikforss, Ö., & Alexander, L. (2007). Rethinking communication in construction. *Journal of Information Technology in Construction, 12,* 337–346.

Xie, X. (2002). *Communications in construction design.* Ph.D. thesis, Loughborough University, Loughborough.

Xie, X., Thorpe, A., & Baldwin, A. N. (2000). A survey of communication issues in construction design. In A. Akintoye (Ed.), *16th Annual ARCOM Conference*, 6–8 September 2000, Glasgow Caledonian University. Association of Researchers in Construction Management (Vol. 2, pp. 771–780).

Xie, C., Wu, D., Luo, J., & Hu, X. (2010). A case study of multi-team communications in construction design under supply chain partnering. *Supply Chain Management: An International Journal, 15*(5), 363–370.

Xue, X., Li, X., Shen, Q., & Wang, Y. (2005). An agent-based framework for supply chain coordination in construction. *Automation in Construction, 14*, 413–430.

Xue, X., Shen, Q., & Ren, Z. (2010). Critical review of collaborative working in construction projects: Business environment and human behaviors. *Journal of Management in Engineering, 26*(4), 196–208.

Xue, X. L., Wang, Y. W., Shen, Q. P., & Yu, X. G. (2007). Coordination mechanisms for construction supply chain management in the Internet environment. *International Journal of Project Management, 25*(2), 150–157.

Chapter 7 Understanding Communication Effectiveness

Abstract Effectiveness of communication among project teams in project delivery has become increasingly important in the wake of growing technical and organizational complexity of construction projects and era of digitalization. Against this, project teams have explored several tools and media as a means of improving communication effectiveness among project teams. However, there are varied and somewhat unknown performance of these tools and media something which is proving to be elusive in practice, especially across various procurement typologies and communication networks. There is growing recognition of the need to understand the communication tools and media performance across various procurement typologies. Currently, focus of most of the studies has been on traditional procurement typologies. This understanding will offer project teams a means of adopting suitable and a balanced mix of rich communication media that suit their interaction and information sharing. With the understanding that no communication media is fit for all context and project typologies, the evidence gives project teams the confidence to adopt fit for purpose media for task and relational interaction and communication in project delivery, especially in various non-traditional procurement typologies. This offers useful suggestions for the construction industry, in particular, on how construction project teams could engage formally in communication interactions and information by using appropriate tools, techniques and communication media.

Keywords Communication effectiveness · Communication media · Communication networks · Project teams

7.1 Introduction

Measuring communication effectiveness or performance is considered a prelude towards communication improvement actions. However, like the broad nature and application of communication, the understanding, measurement and assessment of communication effectiveness and performance has been variedly conceptualized and operationalized in many contexts. In construction context, effective communication measurement has been conceptualized in various forms. However, the thin line of thought in assessment of communication effectiveness has been to identify

T. E. Kwofie et al., *Effective Construction Project Delivery*,
https://doi.org/10.1007/978-3-030-49374-5_7

communication problems and evaluate communication performance among project teams. Arguably, communication effectiveness and communication performance from extant literature refer to the same construct and used interchangeably. Against this, it is important to establish the understanding of communication effectiveness and approach to measuring it.

Given the importance of effective communication of information among the various participants across the life cycle phases, the generation of this knowledge will be beneficial in communication management and strategies that can optimize communication models and styles adopted in construction project environment.

7.2 The Measure of Communication Performance in Construction

The theoretical underpinnings of communication performance or effectiveness lie within the context of task-based and social-based interactions and information sharing in interpersonal and organizational communication. From the field of construction communication, initial efforts by way of research conceptualized communication performance as a means of findings ways of improving information exchange and communication by assessing the problems in the communication, evaluating information flow variables and the communication network variables (Roberts and O'Reilly 1974; Roberto 1997; Guevara and Boyer 1981; Hunter 1993; Mcquail and Windahl 1993; Thomas et al. 1998).

From these studies, the conception of communication performance was the measure of level of distortion, gatekeeping, overload and underload in the communication among the team (Guevara and Boyer 1981; Mcquail and Windahl 1993; Thomas et al. 1998, 1999). The theoretical acceptance of Thomas et al. (1999) is embedded in the fact that these communication variables embrace both the task functional and social theories that should underpin communication improvement paradigms espoused by Marshall-Ponting and Aouad (2005), especially in construction communication and interactions. This indicates that Thomas et al. (1999) has the propensity to identify emerging communication patterns, evaluate information flow and measure the strength of network ties in complex relational communication networks and information sharing (Murray et al. 2000; Xie et al. 2010; Tran et al. 2017).

In this regard, the identification and measurement of critical communication variables was deemed as an important step in improving team communications in construction project delivery (Thomas et al. 1998). The Construction Industry Institute's (CII) research in the USA became a giant progress in a comprehensive evaluation of communication performance by identifying such variables for team communication. The CII categorized the variables in communication performance measures as accuracy, procedures, barriers, understanding, timeliness and completeness of the communication. These variables were used to develop a diagnostic assessment tool "Compass" for improving team communications. The theoretical support of the tool was that, by using the Compass tool, project communication effectiveness can be both measured and monitored, and problem areas can be identified which can be a good resource towards communication improvement efforts (Thomas et al. 1998, 1999; Tran et al. 2017).

From then, the Compass became a vital resource for evaluating communication performance or effectiveness in many studies (Mead. 1999; Xie et al. 2000; Xie 2002; Liu 2009; Xie et al. 2010; Kwofie et al. 2015; Kwofie 2015; Kwofie et al. 2016). From a synthesis and critique of the tool provided by Mead (1999), it was clear that there was the need to expand the questions in the evaluation, especially in the measure of barriers to communication among the team. This was necessary to aid in identifying specific communication barriers including information flow.

From the network variables perspective, the use of social network analysis (SNA) has the primary objective of obtaining high level descriptions and underpinings of a social communication network through an analysis of relational data (Garton et al. 1997). Following this, various computer programmes have also emerged designated to analyzing structure of social networks. Notable among these were the UCINET IV and KrackPlot (Xie 2002; Xie et al. 2010). In the area of network analysis, roles and centrality formed the nucleus of the evaluation (Tichy and Fombrun 1979; Wasserman and Faust 1994). From a broad spectrum of communication performance assessment following the information flow variables, communication variables and the network factors, the measurement is operationalized as in Fig. 7.1.

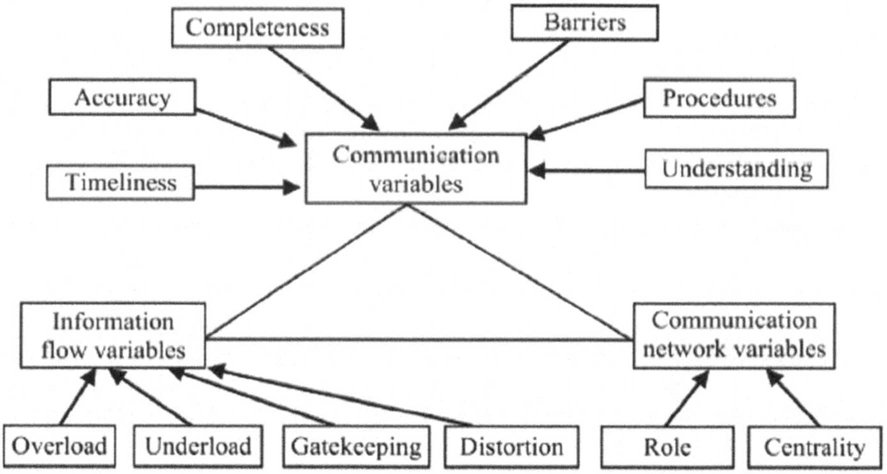

Fig. 7.1 Communication variables and the network factors. *Source* Thomas et al. (1998), Thomas et al. (1999), Xie et al. (2000, 2010)

By way of explanations, the information flow, communication variables and network analysis factors are presented thus.

These variables have formed the communication performance indicators in measuring communication effectiveness in construction communication (Thomas et al. 1998, 1999; Mead 1999; Xie et al. 2000, 2010; Kwofie 2015; Kwofie et al. 2015). In recent times, the dominant approach to communication performance in construction is to explore the influence of communication factors on the communication performance or given attributions to the communication outcome among project teams (see Liu 2009; Xie et al. 2010; Kwofie et al. 2015, 2016; Kwofie et al. 2019a, b; Kwofie et al. 2017).

The evaluation of communication effectiveness in non-traditional procurement involved three facets. These were assessment of the communication performance in typologies of non-traditional procurements, performance of the typologies of communication medium and performance related to typologies of information shared among the team. By this approach, a comprehensive understanding of the communication effectiveness in non-traditional procurement is gained.

7.3 Assessment of Communication Performance in Non-traditional Procurement

It is accepted that communication performance exhibits some complexities in its description as well as its measure. A study was conducted to evaluate the level of communication performance among project teams in various forms of non-traditional procurement systems being used in various project deliveries in South Africa (See Kwofie et al. 2019a, b). For consistency with practice and theoretical validity, the communication performance indicator variables extensively used in measuring communication performance was adapted. For example Kwofie (2015) and Liu (2009) had previously tested the suitability of these variables (Table 7.1) in communication performance assessment in the construction project delivery in various contexts.

The survey was conducted for a period of four weeks through an online portal. Respondents who were identified on these projects were invited via email to respond on a Google form. The forms of non-traditional procurement listed were partnering, project alliancing, integrated project design and public–private partnership (PPP) arrangements. PPPs are generally deemed as a varied form of procurement arrangement distinct from the partnering arrangement (see Cui et al. 2018; Kwofie et al. 2016) hence the separation PPP from project alliance in the assessment. The respondents were invited to draw on their experience from the non-traditional procurement systems used on the projects in which they were involved and indicate the frequency of the communication performance indicators occurring in the communication with other parties in their communication and information sharing in the delivery of the project. The five-point Likert scale was interpreted as the following: 5 = very frequent, 4 = frequent, 3 = occasionally, 2 = rarely, 1 = never.

Table 7.1 Explanation of variables in communication performance

Variables	Explanation
Communication	
Accuracy	Expresses the frequency of conflicting instructions, poor communications and lack of coordination
Barriers	Expresses the interpersonal, accessibility, logistic, etc., barriers interfering with communication between participants
Procedures	Existence, use and effectiveness of formally defined procedures outlining scope, methods, communication strategy, etc.
Completeness	Adequacy or less of relevant information received
Timeliness	Timeliness of information received including design and schedule changes
Understanding	Understanding of information received or expected from other team participants
Information flow	
Distortion	Transformation of the meaning of a message by changing its content
Overload	Receiving more information than that can be utilized or processed, from an individual or system leading to breakdown
Underload	Receiving insufficient information to make decisions from individual or system
Gatekeeping	Controlling messages flowing through a communication channel by positioned individual or group
Communication network	
Role	The participant components of any communication network. Individuals can fulfil several different roles within a project network according to frequency of communication (Tichy and Fombrun 1979)
Centrality	Measures the importance or prominence of an individual in a social network. Through the analysis of the strength of connections with other members of the communication network

Sources Tichy and Fombrun (1979), Guevara and Boyer (1981), Wasserman and Faust (1994), Garton et al. (1997), Thomas et al. (1998, 1999), Mead (1999), Xie et al. (2000), Liu (2009), Xie et al. (2010), Kwofie et al. (2015, 2016, 2017)

There was a total of 169 responses received forming a response rate of 59%. From the argument espoused by Takim et al. (2004), the 59% could be deemed as an acceptable response rate compared to the indicated 20–30% as the normal acceptable response rate for a questionnaire survey, especially in construction management research. Aggregation of the respondents revealed project managers (18%), architects (12%), contract/construction managers (10%), engineers (12%), quantity surveyors (29%), contractors (10%) and sub-contractors (6%) and suppliers (3%). A breakdown of the typology of non-traditional procurements involved in revealed that PA, IPD, partnering and PPP were the notable forms identified by respondents as used in the projects in which they were involved. Their experiences revealed that 32% had between 0 and 5 years, 41% with 6–10 years, 15% had about 11–15 years of experience and those above 15 years were 12%.

Mean scores, inter-rater agreement and Mann–Whitney U test results are presented in Table 7.2. From Table 7.2, all the results indicate that all the values for the standard mean error were relatively small (less than 1.0). The relatively small standard mean errors suggest that the sample mean is similar to the population mean and so the sample is likely to be an accurate reflection of the population (Field 2009; Motulsky 2005). From the standard mean error results (see Table 7.2), it can incontrovertibly be stated that the sample chosen for the study adequately reflects the population, and thus, conclusions and generalization from the findings are more likely to be accurate and respectable. Additionally, it can also be seen that inter-rater agreement index assessment (rWG) in Table 7.2 was all 0.5. An inter-rater agreement index assessment value of above 0.5 denotes a substantial to a near perfect agreement (Cohen et al. 2009). Hence, the results can be described as substantial to near perfect agreement, suggesting a significant consensus among the respondents in their ratings of the frequency of the communication performance. The inherent deduction from this is that the aggregated mean ratings given by the respondents can be deemed as being a good and acceptable representation of their experience and professional judgment of the communication performance in non-traditional procurement.

The Mann–Whitney U test using the conventional significance level of 95% was conducted to assess whether any significant differences exist in the frequency ratings of the communication performance indicators in the two main domains of interpersonal and organizational communications in the projects they were involved in their typology of non-traditional procurement systems (see Liu and Shen 2008; Motulsky 2005). In this regard, a communication performance attribute of a p value score less than 0.05 is an indication of a significant difference in the perception and ratings of the respondents in interpersonal and organizational communication of in their tasked-based and social-based interactions.

The Mann–Whitney U test was conducted with the hypothetical axiom that there are no significant differences in the communication performance in respect of the interpersonal and organizational communication among the teams in the adopted non-traditional procurement system. From the results in Table 7.2, one could note that only two variables under accuracies and gatekeeping had their p values less than 0.05 implying that generally the projects teams engaged in the various non-traditional procurement systems experienced similar communication performance in their interpersonal and organizational communication. A plausible explanation to this is that, it is more likely that the social and project environment and managerial systems yield consistent or similar degrees of communication performance in interpersonal and organizational communication in non-traditional procurement systems adopted in project delivery. These myriads of tests were aimed at enhancing rigor as well as giving credence to the conclusions and generalization of the findings. The following sections discuss the findings by focusing on the main communication performance indicators (themes).

Table 7.2 Results of means scores, inter-rater agreement and Mann–Whitney's test on communication performance indicators

Indicators	Explanatory variables (communication performance indicators)	P		PA		IPD		PPP	
		Mean	Mean	Mean	Mean	SME	Inter-rater	Mann–Whitney U	Exact sig. (two-tailed)
Accuracies	Receiving conflicting information from team participants	4.31	4.06	4.33	4.19	0.061	0.63	269.0	0.867
	Lack of consistency in communicated information leading to ineffective coordination among project team	3.76	3.94	3.54	2.18	0.079	0.69	134.3	0.003*
	Lack of conciseness in communicated information among the project team	3.46	3.09	3.22	3.01	0.082	0.81	242.5	0.387
Timeliness	Late delivery of needed communicated information	2.06	1.99	2.13	2.01	0.072	0.88	196.4	0.060
Distortions	Lack of coherency in meaning of communicated information	3.27	3.09	3.01	3.47	0.102	0.96	271.0	0.774
	Notable distortions in content of communicated information	2.09	2.28	2.42	2.06	0.087	0.73	265.2	0.804
	Lack of clarity in communicated information resulting in inconsistent interpretations	3.01	3.00	3.13	3.07	0.079	0.91	193.3	0.095
Barriers	Difficulty in access to communicated and shared information from channels	1.04	1.46	1.08	1.01	0.082	0.68	227.5	0.251
	Inefficient dissemination of information among project team in channels	1.37	1.30	1.47	1.28	0.098	0.83	218.0	0.172
Underloading	Receiving less information than expected from team participants for tasks	1.93	1.62	1.80	1.55	0.061	0.75	145.2	0.372

(continued)

Table 7.2 (continued)

Indicators	Explanatory variables (communication performance indicators)	P		PA		IPD		PPP	
		Mean	Mean	Mean	Mean	SME	Inter-rater	Mann–Whitney U	Exact sig. (two-tailed)
Overloading	Receiving more information than necessary for the tasks than expected	1.74	1.69	1.62	1.92	0.048	0.69	194.8	0.116
Understanding	Misunderstanding communicated information	2.18	2.00	2.39	2.08	0.037	0.65	199.5	0.126
Gate keeping	Withholding of part of the information by the one who controls communication	2.22	2.07	1.97	4.01	0.069	0.86	282.5	0.012*
	Withholding of whole of the information by the one who controls communication	2.19	2.06	2.03	2.11	0.046	0.63	247.0	0.580
Procedure	Inefficient disseminating protocols relating information sharing among teams	3.94	4.01	3.98	4.44	0.028	0.83	262.5	0.802
	Lack of clearly defined roles and responsibilities among members of the team	1.08	1.02	1.17	1.07	0.059	0.89	266.0	0.866

Note *Significant level (0.05) 95% confidence interval (*P* partnering; *PA* project alliance; *IPD* integrated project delivery; *PPP* public–private partnership)

7.3.1 Accuracy in the Communication Among the Team

Three variables measured the level of accuracy in the communication performance among the project teams across the typologies of non-traditional procurement systems adopted in project delivery. From the results, it could be deduced that by approximating the mean score of each criterion to the nearest whole number to conform to the scale used and aid ease of interpretation, the score on *receiving conflicting information from team participants* could be described as being frequent among the team. In the case of *lack of consistency in communicated information leading to ineffective coordination among project team,* the score could be described as also being frequent across all typologies except in PPP delivery. However experiencing lack of conciseness in communicated information among the project team among the team in their interpersonal and organizational communication was occasionally present. The communication performance in accuracies of the interactions seems to be consistent with the revelations by Xie et al. (2010) stating that sharing of inaccurate information is a major notable communication performance problem in construction partnering supply chain. In comparison with the findings by Liu (2009), it can be said that inaccuracies in communication among project team is a phenomenon experienced in both conventional and non-traditional procurement systems.

However, with the incidence of occasional performance in *lack of consistency in communicated information leading to ineffective coordination among project team* it can be asserted that non-traditional procurement systems offer a chance for an improved communication performance in accuracy of communication compared to traditional procurement systems. Likewise, it can further be stated that PPP project environment offers different social and project environment and managerial systems that experience unique communication performance in the consistency in communicated information leading to effective coordination among project team compared to partnering, IPD and project alliance. The empirical exposition given by Xie et al. (2000) in accuracy in the communication performance in construction projects proved that it is one the most important factors contributing to team performance and managerial effectiveness. However, the high incidence of inaccurate communication in construction project delivery is attributed to human factors such as poor coordination, conflicting and poor communication skills (Guevara and Boyer 1981; Thomas et al. 1998; Xie et al. 2010; Tran et al. 2017). From this, it can be posited that efforts aimed at improving communication skills that can help improve accuracy in communication in the construction project delivery so as to optimize the benefits inherent from non-traditional procurement systems.

7.3.2 *Timeliness and Procedure in the Delivery of the Communication Information*

The score on *late delivery of needed communicated information* can be described as being rare among the teams across the various non-traditional procurement systems adopted. In the case of following procedure in the delivery of communicated information, *inefficient disseminating protocols relating information sharing among teams* emerged as being frequent whereas *lack of clearly defined roles and responsibilities among members of the team* never occurred. According to Xie et al. (2000), late delivery of required communicated project-related information has persistently been a highly ranked frequent communication problems in the construction project delivery in traditional procurements. In contrast to non-traditional procurement, it can be stated unequivocally that the non-adversarial nature coupled with non-rigid contractual arrangements allow participants to freely share information and on time thus leading to better performance in timeliness compared to traditional procurement systems. In the case or performance of the procedure and protocols in team communication, the study by Xie et al. (2000) revealed that there is high incidence of bureaucratic administrative procedures, rigid contractual arrangement and protocols, communication channels and links, information prioritization and role conflict as the root causes of untimely delivery of construction communicated information in traditional procurements as well as procedure consistency.

From the results of untimely communication and lack of clearly defined roles and responsibilities among members of the team being rare never occurring is an indication of better administrative procedures, flexible contractual arrangement and good role definitions well as efficient communication channels and links in IPD, PPP, partnering and project alliance as intimated by Kwofie et al. (2019b). However in the case of *inefficient disseminating protocols relating information sharing among teams,* it can be said that both traditional and non-traditional procurement systems suffer from inefficient disseminating protocols. Thus, even though Chen and Manley (2014) revealed that non-traditional procurement tends to adopt flexible non-rigid formal processes, procedures and flexible governance structures compared to the traditional procurement, this attribute does not significantly facilitate sharing of information in the area of procedures and protocols. In contrast, it can also be construed that non-traditional procurement systems social and organizational environment reduce the notable complex communication environment associated with traditional procurement systems, and thus offer enhance communication performance in timeliness of the shared information and clarity in defined roles and responsibilities among members of the team.

7.3.3 Performance in Relation to Distortions in the Communication

The results on the measure of distortions in the communication across the PPP, PA, IPD and partnering can be described as being rare occasional (see Table 7.2). This suggests that distortions associated with information composed and shared in the project communication in non-traditional procurement was low. Hence, there was good communication performance in coherency, clarity and uniform content in the communicated information among project teams. This revelation is rather a sharp contrast to Mead (1999) where distorted information was common to project team and contributed to conflict in nine companies using traditional procurement systems. However, the nature of performance in distortion of information as reported by the study seems to be consistent with Xie et al. (2000) and Liu (2009) wherein distorted communicated information was noted as being a rare communication problem among project teams at the design and construction stages in traditional procurement systems. The varying findings from Mead (1999), Xie et al. (2000) and Liu (2009) could suggest that distortions in communication among construction project teams is more than the influence of the procurement systems. Given that sharing project-related information devoid of distortions is deemed crucial irrespective of the procurement systems (Murray et al. 2000; Shohet and Frydman 2003; Alshawi and Ingirige 2003; Tran et al. 2017), distortion-free communication should be a must to engender success.

7.3.4 Level of Barriers Experienced in the Team Communication

The works of Kwofie (2015), Xie et al. (2000, 2010), Alshawi and Ingirige (2003), Liu (2009) and Dawood et al. (2002) all agree that both human and technological barriers have been ranked high among the major blocks in access to the shared information among project teams, chiefly at the design and construction stages in traditional procurement systems. From the results in Table 7.2, there is the incidence of rare difficulty in access to the communicated and shared information from channels and inefficient dissemination of information among project teams in communication networks and channels. Shen (1992) and Dawood et al. (2002) revealed that access to shared information was difficult, and this was primarily attributed to poor presentation or hierarchy barriers. The contrast presented by the study is an indication that non-traditional procurement allows non-significant effect of human and technological barriers which are common in traditional procurement systems.

7.3.5 Completeness in the Communication

Completeness of the communication was measured by the level of underload and overload in the shared information. In Table 7.2, the results indicate rare incidence of *receiving more information than necessary for the tasks than expected* and *receiving less information than expected from team participants for tasks.* Xie et al. (2000) and Thomas et al. (1998) intimated that the need to ensure completeness in communicated information has repeatedly been cited as a sine qua non to task function, decision-making and team performance. The situation where project suffer from less information (underload) or more than needed information (overload) significantly affects task functions, social cohesion and team performance (Liu and Shen 2008; Alshawi and Ingirige 2003; Tran et al. 2017). Liu (2009) provided a practical insight on two studies in Chnia and Hong Kong revealing that receiving less or more than enough information to make decisions and implement work was a very frequent phenomenon in projects in both countries. This revelation confirmed an earlier study by Thomas et al. (1998). This makes the revelation by Xie et al. (2010) that information overload is a major issue in partnership construction supply chain very interesting, requiring an in-depth assessment.

7.3.6 Understanding of the Communication

Xie (2002), Mead (1999) and Thomas et al. (1998) affirmed that understanding communicated information is the most significant aspect of communication performance, and thus, when shared information is not well understood, the main objective of communication is always lost. In assessing the performance of understanding in non-traditional procurement typologies, an approximated mean score of 2.0 is a testament to a rare incident of *misunderstanding communicated information* in the adopted non-traditional procurement systems. The findings suggest that misunderstanding was a rare occurrence among the project teams in the adopted non-traditional procurement systems. This tend to support the revelations by Xie et al. (2010) that misunderstanding is not common in project alliance and partnering arrangements in project delivery. However in traditional procurement systems, misunderstanding information expectations and requirements between project team participants was a notable problem, especially at the construction stage. From this it can be plausibly explained that the well-acknowledged informal nature, governance and environment in non-traditional procurement afford project team participants to seek further insights and enquiries that could enhance the understanding of the information compare to the notable adversarial nature of traditional procurements. These attributes of non-traditional procurement systems tend to alleviate the obvious adversarial posture and environment often encountered in traditional procurement systems and thus enhance the desire and willing in sharing information and pursuit of seeking clarification over and over to improve understanding.

7.3.7 Gate Keeping as a Communication Challenge

Gatekeeping as a construction communication problem refers to a situation wherein persons acting as gatekeepers regulate the flow of information among project teams (Thomas et al. 1998; Liu 2009; Xie et al. 2010). This phenomenon has overly been cited as being a common incident in traditional procurements (Thomas et al. 1998; Liu 2009; Xie et al. 2010). From the results reported here, it can be deduced that gatekeeping as a communication problem was rare in occurrence in non-traditional communication. This emergence suggests that non-traditional procurement systems offer a free communication environment which does not warrant any control of the communication. This could support the finding by Pryke (2004) that perhaps the free social environment in non-traditional procurement is a boost for enhanced willingness to share project-related information compared to traditional procurements.

The findings herein presented have sought to bridge the knowledge gap with respect to the communication performance among project teams in typologies of non-traditional procurement systems. An empirical knowledge and understanding of frequency of the occurrence of communication performance issues or problems in non-traditional procurement systems, focusing on the interpersonal and organizational communication among the project teams, has been elucidated in depth. Findings generally affirm a better performance of communication in non-traditional procurement systems. Though some of the findings concurred with already known communication performance issues in traditional procurements systems, there are significant revelations affirming that indeed non-traditional procurements systems are unique and indeed exhibit communication environment different from traditional procurement models. This in mind must stimulate an agenda to adopt bestspoke communication models and strategies that fit the features of non-traditional procurement systems so as to optimize it benefit inherent from effective interpersonal and organizational communication in project delivery.

The findings also seem to corroborate the fact that the free communication environment experienced in non-traditional procurement systems could plausibly be the explanation to the varied differences in communication performance in relation to gatekeeping, procedure and protocols compared to traditional procurement systems. With the increasing adoption of non-procurement systems over traditional procurement methods, coupled with the widespread need to improve construction communication (see Dainty et al. 2006; Emmitt and Gorse 2003; Xie et al. 2010), communication performance problems revealed here deserve attention. It can be argued that it is only when communication performance issues relating to any context reveal that effective measures can be pursued towards improvement. The findings illustrated can be considered very vital and significant in efforts on communication planning, management, protocols to curtail these frequent communication challenges. These are indeed necessary aspects of multi-team communication towards enhanced integrated teams, team and task performance and overall project performance. Overall, the findings offer practical implications to practitioners and stakeholders in their quest to significantly improve communication performance in non-traditional procurement in order to optimize its benefits.

7.4 Performance of Communication Medium in Non-traditional Procurement

Ensuring effective communication performance among the project team has become an increasingly important factor across the life cycle phases due to the growing technical and organizational complexity of construction projects and the urge for a more integrated and collaborative working process in project delivery. The adoption of the most suitable and appropriate communication medium for the team communication, task-based and social interactions and information sharing has been seen as one way of improving the communication performance of project design and construction teams (Otter and Emmitt 2007; Emmitt and Gorse 2007; Liu 2009; Tran et al. 2017).

The construction industry has currently used numerous communication media, tools and strategies in the communication and information sharing among the project teams. Each communication medium is said to have its own strengths and weaknesses, and there is no single delivery medium appropriate for every communication situation, project environment, information typology, procurement system, project and organizational complexities to ensure effective communication among the project team in the construction project delivery (Otter and Emmitt 2007; Manninen 2003). For any communication situation and requirements on any particular project, teams and stakeholders need to examine the merits and understanding of the performance of communication media and other relevant factors to determine how they suit and align with the attributes of the project environment, procurement systems in order to engender the needed communication performance outcome.

Inherent from the many documented uniqueness of non-traditional procurement systems, there is the need to gain theoretical and practical understanding of the nature of performance of various forms of communication media that can inform the decision on the choice and suitability in communication, interactions and information sharing among project teams. Likewise, there is also growing recognition of the need to understand the communication media alongside the needs of the project participants and how they impact on the communicate effectiveness in the communication process within project teams (Emmitt and Gorse 2007; Otter 2005).

Typically, construction project team who communicate and share project information are multidisciplinary in nature coming from different organizations, which have different organizational cultures relying on varied communication media and information systems to undertake their communication task (Robbins 2001; Xue et al. 2010; Xie et al. 2010). The project team participants may also show varied form of skills and competencies in the use and adoption of these communication media and systems as well as preferences for specific means of communication and context

(Xie et al. 2010; Robbins 2001; Gorse 2002). Otter and Emmitt (2007) revealed that the effectiveness of the communication among the project team highly dependent on two inter-related factors which the preferences for using specific communication media is one that is crucial in addition to the competences of team members in the use of the media.

Construction project team communication is most effective when all members contribute using the available communication media in the same way in a balanced mix of synchronous and asynchronous communication (Otter and Emmitt 2007; Gorse 2002; Gorse and Emmitt 2007). Otter and Emmitt (2007), Liu (2009) and Gorse and Emmitt (2007) revealed that construction communication make use of host of communication media and means that includes Dialogue, group meeting, informal meeting, telephone, facsimile, postal mail, project dossier, email message, messenger service, video conference, Outlook calendar, computer network (intranet) and project Website. However, in recent times, the digital age has made advances to develop other platforms such as BIM and other 3D computer-aided forms for information sharing where issues emanating from the context of information sharing and behaviours still exist (Ibrahim et al. 2019; El-Saboni et al. 2009). From an objective survey of 162 responses using questionnaire, the results of the nature of performance of typologies of communication media in project delivery using partnering, project alliance, IPD and PPP as non-traditional projects system is presented in Table 7.3. The criteria of

Table 7.3 Effectiveness and utility of the communication media in non-traditional procurements

Communication Media	P			PA			IPD			PPP		
	U	E	F	U	E	F	U	E	F	U	E	F
Face-to-face formal dialogues	1.37	4.12	3.01	0.94	4.07	4.34	1.85	3.56	1.92	0.90	3.31	3.69
Project meetings	1.00	4.88	4.86	0.92	4.24	4.62	1.02	4.11	4.04	0.82	3.93	4.77
Telephone	0.57	2.02	3.57	1.01	4.04	4.01	0.38	1.49	3.96	0.46	2.04	4.46
Video conference	1.87	3.56	1.90	0.86	2.28	2.64	0.96	1.03	1.07	0.46	1.92	4.13
Postal mail	0.54	1.12	2.08	3.74	3.93	1.05	3.50	4.13	1.18	1.44	2.33	1.62
Facsimile	0.66	2.08	3.15	0.97	2.00	2.07	1.05	2.06	1.96	1.62	1.67	1.03
Email	1.08	4.67	4.32	1.05	4.86	4.63	0.86	4.09	4.78	0.94	4.59	4.88
Intranet	0.84	1.23	1.46	1.95	3.93	2.02	0.64	2.60	4.21	0.76	2.99	3.91
Outlook agenda	0.45	2.04	4.53	0.78	2.07	2.64	0.49	1.13	2.29	1.25	1.32	1.06
Project Website	0.99	4.02	4.11	0.91	3.81	4.19	1.02	3.09	3.01	1.85	3.68	1.99
BIM (3D platforms)	1.25	3.81	3.06	0.94	4.08	4.33	1.06	4.28	4.05	2.05	3.84	1.87

P Partnering; *PA* project alliance; *IPD* integrated project delivery; *PPP* public–private partnership; *U* utility; *E* effectiveness; *F* frequency of use frequency of use of the media, effectiveness and utility were used. The utility is a measure of the ration of the effectiveness to the frequency of use. This approach is consistent with Liu (2009) and Otter and Emmitt (2007).

The results from Table 7.3 suggest that in partnering, face-to-face formal dialogues, project meetings, video conference, email, project Website and BIM (3D platforms) were the efficient communication media whereas telephone, fascmile, project intranet, Outlook and post mail were found to be less effective in spite of their frequency of usage. From the utility analysis, it was noted that face-to-face formal dialogues, project meetings video conference, email, intranet and BIM (3D platforms) had their utility values more than 1.00 (utility > 1) meaning their effectiveness far exceeded their frequency of usage. This implies that practitioners, stakeholders, project teams can pay more attention to increase their usage given their effectiveness performance. Liu (2009) found telephone, meeting and face-to-face discussion as being more efficient in traditional procurement models. From this, it can be said that the performance of face-to-face,

Formal dialogues, project meetings video conference, email, intranet and BIM (3D platforms) are good and can be used extensively in partnering project arrangements.

In project alliance project delivery, face-to-face formal dialogues, project meetings, telephone, postal mail, project Website and BIM (3D platforms) were found to be efficient in their communication usage. Additionally, email, intranet and telephone were found to have their utility greater than 1.00 meaning, and their effectiveness was greater than their frequency of usage. This offer opportunity for project teams to increase their usage so as to optimize their effectiveness in their communication in project alliance. From integrated project delivery (IPD), face-to-face formal dialogues, project meetings, postal mail, email and BIM (3D platforms) were found to be the most effective communication media. In terms of their utility, they were values were greater than 1.00. This is an indication that the use of these communication media can be increased by the communication network. Telephone, video conference and Outlook agenda were found to be less effective in IPD arrangements. The emergence of email and project meetings seems to be consistent with Xie et al. (2010) in IPD and alliance arrangements.

The results also show that in PPP project communication, face-to-face formal dialogues project meetings, telephone, video conference, email and intranet were the most frequently used communication media. However, face-to-face formal dialogues, project meetings, email, project Website and BIM (3D platforms) were found to be effective. Interestingly, postal mail and Outlook agenda had high utility scores, but they were less effective. This is a sharp contrast to the revelation by Liu (2009) where postal mail and Outlook agenda were found to be effective in project delivery in China in traditional procurement. This suggests that the use of these in PPP must be done cautiously. From a glance, it can be concluded that across the four typologies of non-traditional procurement, face-to-face formal dialogues, project meetings, email, project Website and BIM (3D platforms) were found to be the most effective communication media. The practical implication of this is that project teams can increase their usage so as to optimize their benefits towards effective communication outcome in non-traditional procurement systems.

7.5 Performance Related to Information and Documents in Non-traditional Procurement

Pietroforte (1997) indicated that project teams share information needed to represents the requirements of the building to be constructed and information for the coordination of the activities to realize the requirements of the projects. These information are encoded and sent through a medium where the receivers must correctly decode them to aid the right understanding that will lead to the right actions (Pietroforte 1997, 1999; Shen 1992). Mead (1999) classified construction information into technical information, commercial information and monitoring and control information. Technical information entails drawings, specifications, details and design clarifications. Commercial information on the other hand included delivery schedules, costs and payments, administration terms and conditions (Mead 1999; Shen 1992). Meeting minutes, change orders, status logs, as-built drawings, contract status log, safety information daily logs, project schedules, contract documents and requisitions (Pietroforte 1997; Shen 1992).

Pietroforte (1997) intimated that the transmission of these information (documents) must be weighed alongside the communication media being used. Generally, some of these documents could be communicated using multiple media. For example, the contract documents could be communicated using face-to-face formal dialogue, project meeting, post mail, project Website or email. However and in-depth understanding of the performance of these typologies of information in the various media will be a variable decision information. Using a similar approach by Otter and Emmitt (2007), the performance of the three typologies of information (technical, commercial and monitoring and control) were explored. The markings √, x and 0 were used to rate their performance interpreted as high level, average level and low level, respectively, using ease of use, feedback, interaction, overview, informal, formal and status as the measuring criteria. From Otter and Emmitt (2007) and Pietroforte (1999) "ease of use" refers to the interface of the means with the user(s) is simple and easy; "feedback", direct feedback of the receiver to the sender; "interaction", immediate repeated feedback between sender and receiver; "overview", the information collected is complete and can be viewed in total; "informal", without restrictions or rules; "formal", with restrictions and rules to follow; "status", the status of stored information (new, updated, final).

The assessment was generic to non-traditional procurement without reference to any particular type. The respondents were asked to rate each of the media used in the communication of technical, commercial and monitoring and control information on the seven criteria of effectiveness using high level, average level and low level. The results are given in Table 7.4.

The results show that email, project meetings and project Website are easy to use, offer prompt feedback and high-level status with respect of storage and update. With regard to BIM (3D platforms) which offers high-level feedback and status performance, its ease of use was very low among the team. Project meetings, telephone

Table 7.4 Performance of topology of information and documents in non-traditional procurement systems

Means of communication	Technical, commercial and monitoring and control information						
	Ease of use	Feedback	Interaction	Overview	Informal	Formal	Status
Face-to-face formal dialogues	√	√	√	0	x	x	0
Project meetings	√	√	√	0	0	√	x
Telephone	√	√	√	0	X	x	0
Video conference	x	√	√	0	0	√	o
Postal mail	0	0	0	0	0	√	x
Facsimile	x	0	0	0	0	√	x
Intranet	√	0	x	x	X	X	√
Email	√	x	x	√	√	x	√
Outlook agenda	√	0	0	x	0	x	0
Project Website	√	√	0	x	0	√	√
BIM (3D platforms)	x	√	0	√	0	√	√

√ = High level, x = Average level and 0 = Low level

and Outlook agenda were easy to use but low in status. In the aspect of interactions, overview and formal, project meetings, BIM (3D platforms), email and Postal mail showed varied performance. Overall, it can be said that project meetings, email, project Website and BIM (3D platforms) showed good performance in non-traditional procurement communication in the area of ease of use, feedback, interaction, formal and status of stored information (new, updated, final).

7.6 Summary

Understanding of the nature of communication performance in non-traditional procurement, communication media and typologies of information and documents is considered important for choosing the right media and tools that suit the technical and organizational complexity of construction project environment. This chapter has an in-depth insight into this aspect of communication in non-traditional procurement that is crucial towards the choice of communication media and strategies that can optimize effective communication in non-traditional procurement. Beyond this, other factors that can influence effective communication in non-traditional procurement must be pursued as a necessity.

References

Alshawi, M., & Ingirige, B. (2003). Web-enabled project management: An emerging paradigm. *Automation in Construction, 12*(4), 349–364.

Chen, L., & Manley, K. (2014). Validation of an instrument to measure governance and performance on collaborative infrastructure projects. *Journal of Construction Engineering and Management, 140*(5), 04014006. https://doi.org/10.1061/(ASCE)CO.1943-7862.0000834.

Cohen, A., Doveh, E., & Nahum-Shani, I. (2009). Testing agreement for multi-item scales with the indices rwg(j) and ADm(j). *Organizational Research Methods, 12*(1), 148–164.

Cui, C., Liu, Y., Hope, A., & Wang, J. (2018). Review of studies on the public–private partnerships (PPP) for infrastructure projects. *International Journal of Project Management, 36*(5), 773–794.

Dainty, A., Moore, D., & Murray, M. (2006). *Communication in Construction-Theory and practice*. 2 Park Square, Milton Park, Abingdon, Oxon: Taylor & Francis.

Dawood, N., Akinsola, A., & Hobbs, B. (2002). Development of automated communication of system for managing site information using internet technology. *Automation in Construction, 11*(3), 552–572.

El-Saboni, M., Aouad, G., & Sabouni, A. (2009). Electronic communication systems effects on the success of construction projects in United Arab Emirates. *Advanced Engineering Informatics, 23*(1), 130–138.

Emmitt, S., & Gorse, C. A. (2003). *Construction communication*. Oxford: Blackwell Publishing.

Emmitt, S., & Gorse, C. A. (2007). *Communication in construction teams*. London: Taylor & Francis.

Field, A. (2009). *Discovering statistics using SPSS for windows* (3rd ed.). London: London Sage Publications.

Garton, L., Haythornthwaite, C., & Wellman, B. (1997). Studying online social networks. *Journal of Computer-mediated Communication, 3*(1).

Gorse, C.A. (2002). *Effective interpersonal communication and group interaction during construction management and design team meetings*. Ph.D. thesis, University of Leicester.

Gorse, C. A., & Emmitt, S. (2007). Communication behaviour during management and design team meetings: A comparison of group interaction. *Construction Management and Economics, 25*, 1197–1213.

Guevara, J. M., & Boyer, L. T. (1981). Communication problems within construction. *Journal of Construction Engineering, American Society of Civil Engineers, 107*(4), 552–557.

Hunter, J. M. G. (1993). *Communications in construction*. Unpublished M.Sc. Dissertation, Department of Civil and Building Engineering, Loughborough University, UK.

Ibrahim, C. K. I. C., Sabri, N. A. M., Belayutham, S., & Mahamadu, A. (2019). Exploring behavioural factors for information sharing in BIM projects in the Malaysian construction industry. *Built Environment Project and Asset Management, 9*(1), 15–28.

Kwofie, E. (2015). *Contribution of unique features of mass housing projects to project team communication performance*. unpublished PhD thesis, Kwame Nkrumah University of Science and Technology, Department of Building Technology, Kumasi.

Kwofie, T. E., Adinyira, E., & Fugar, F. (2017). Theoretical and practical implications for engendering project team communication effectiveness in mass housing project delivery in Ghana. *Journal of Engineering, Design and Technology, 15*(6), 826–844.

Kwofie, T. E., Aigbavboa, C. O., & Machethe, S. O. (2019a). Nature of communication performance in non-traditional procurements in South Africa. *Engineering, Construction and Architectural Management, 26*(10), 2264–2288.

Kwofie, T. E., Aigbavboa, C. O., & Thwala, W. D. (2019b). Communication performance challenges in PPP projects: Cases of Ghana and South Africa. *Built Environment Project and Asset Management, 9*(5), 628–641.

Kwofie, T. E., Fugar, F., & Adinyira, E. (2015). Contribution of multiple construction site management features to project team communication effectiveness: The case of mass housing projects. *Engineering Project Organization Journal, 5*, 180–193.

Kwofie, T. E., Fugar, F., & Adinyira, E. (2016). Modelling the effect of housing design unit contract packaging on mass housing project team communication performance. *Journal of Construction in Developing Countries, 21*(1), 35–50.
Liu, Y. (2009). *Critical Factors for Managing Project Team Communication at the construction stage*. Ph.D. Thesis. Submitted to the Polytechnic University of Hong Kong.
Liu, Y., & Shen, L. (2008). Factors influencing project team communication and its management. In *Proceedings, CRIOCM 2008 International Research Symposium on Advancement of Construction Management and Real Estate* (pp. 219–223). Beijing, China, 31 Oct.–3 Nov. 2008.
Manninen, T. (2003). *Conceptual, communicative and pragmatic aspects of interaction forms-rich interaction model for collaborative virtual environments*. In *The Proceedings of Computer Animation and Social Agents (CASA) Conference* (pp. 168–174). IEEE Computer Society Press.
Marshall-Ponting, A. J., & Aouad, G. (2005). An nD modelling approach to improve communication process for construction. *Automation in Construction, 14*(3), 311–321.
Mcquail, D., & Windahl, S. (1993). *Communication models*. London and New York: Longman.
Mead, S. T. (1999). *Communication effectiveness in intranet based construction projects*. Unpublished Ph. D thesis, Department of Civil and building engineering, Loughborough, University, Loughborough, UK.
Motulsky, H. J. (2005). *Prism 4 statistics guide–statistical analyses for laboratory and clinical researchers*. San Diego, CA: GraphPad Software Inc.
Murray, M. D., Tookey, J. E., Langford, D. A., Hardcastle, C. (2000). Project communication variables: a comparative study of US and UK construction industry perceptions. In *ARCOM Conference Proceedings* (pp. 813–22), Glasgow, UK.
Otter, A. F. H. J. (2005). *Design team communication using a project website*. PhD thesis, Bouwstenen 98, Technische Universiteit Eindhoven, Eindhoven.
Otter, A., & Emmitt, S. (2007). Exploring effectiveness of team communication: Balancing synchronous and asynchronous communication in design teams. *Engineering, Construction and Architectural Management, 14*(5), 408–419.
Pietroforte, R. (1997). Communication and governance in the building process. *Construction Management and Economics, 15*(1), 71–82.
Pietroforte, R. (1999). *Communication and information in the building delivery process*. PhD thesis, Massachusetts Institute of Technology, Cambridge, MA, USA.
Pryke, S. D. (2004). Analysing construction project coalitions: Exploring the application of social network analysis. *Construction Management and Economics, 22*(8), 787–797.
Robbins, S. P. (2001). *Organizational behavior*. Englewood Cliffs, NJ: Prentice-Hall.
Roberto, P. (1997). Communication and governance in the building process. *Construction Management and Economics, 15*, 71–82.
Roberts, K. H., & O'Reilly, C. A. (1974). Measuring organizational communication. *Journal of Applied Psychology, 59*, 321–326.
Shen, L. Y. (1992). *Information management in construction companies*. Hong Kong: Hong Kong Polytechnic University, Department of Building and Surveying.
Shohet, I. M., & Frydman, S. (2003). Communication patterns in construction at construction manager level. *Journal of Construction Engineering and Management, 129*(5), 570–577.
Takim, R., Akintoye, A., & Kelly, J. (2004). Analysis of performance measurement in the Malaysian construction industry. In S. O. Ogunlana, C. Chareonngam, P. Herabet, & B. H. W. Hadikusumo (Eds.), *Proceeding of Globalization and Construction* (pp. 533–546), AIT Conference Centre, Bangkok, CIB, Rotterdam.
Thomas, S. R., Tucker, R., & Kelly, W. (1998). Critical communications variables. *Journal of Construction Engineering Management, 129*(1), 58–66.
Thomas, S. R., Tucker, R., & Kelly, W. (1999). Compass: An assessment tool for improving project team communications. *Project Management Journal, 30*(4), 15–23.
Tichy, N. M., & Fombrun, C. (1979). Network analysis in organisational settings. *Human Relations, 32*, 923–965.

Tran, D. Q., Nguyen, L. D., & Faught, A. (2017). Examination of communication processes in design-build project delivery in building construction. *Engineering, Construction and Architectural Management, 24*(6), 1319–1336.

Wasserman, S., & Faust, K. (1994). *Social network analysis; Methods and Applications*. New York: Cambridge University Press.

Xie, X. (2002). *Communications in construction design*. PhD thesis, Loughborough University, Loughborough.

Xie, X., Thorpe, A., & Baldwin, A. N. (2000). A survey of communication issues in construction design. In A. Akintoye (Ed.), *16th Annual ARCOM Conference* (Vol. 2, pp. 771–780), 6–8 September 2000, Glasgow Caledonian University. Association of Researchers in Construction Management.

Xie, C., Wu, D., Luo, J., & Hu, X. (2010). A case study of multi-team communications in construction design under supply chain partnering. *Supply Chain Management: An International Journal, 15*(5), 363–370.

Xue, X., Shen, Q., & Ren, Z. (2010). Critical review of collaborative working in construction projects: Business environment and human behaviors. *Journal of Management in Engineering, 26*(4), 196–208.

Chapter 8
Critical Factors for Managing Non-traditional Procurement Communication

Abstract In the current construction industry, various procurement models adopted necessitates a more integrated, cooperative and collaborative working relying on continuous flow and sharing of project-related information among project teams for decisions and task functions. Ensuring effective communication performance for project delivery is said to be dependent on salient communication factors that must be pursued rigorously. However, existing studies in this domain have persistently focused on tradition procurement models which are entirely different from non-traditional ones. These factors when explored will be valuable for engendering effective communication which is a vital process for collaborative working and integration among project teams in construction project delivery. The communication factors and communication performance approach has become a precursor for identifying the impact of critical factors on communication performance and thus reveals areas for improvement, thus providing the framework for project teams to develop tailored strategies for improving communication performance among project teams in non-traditional procurement models. Through a questionnaire survey, an insight is provided on the critical factors that impact on communication performance across various non-traditional procurement models. By this knowledge, project teams are primed to evolve strategies and tools that are significant to improving communication effectiveness in non-traditional procurement models. This will help participants to communicate more effectively and ultimately improve the quality of construction project outcomes and improvement in processes.

Keywords Communication factors · Non-traditional procurement · Communication performance · Effective communication

8.1 Introduction

Construction project delivery process in various procurements systems is said to be technically and organizationally complex requiring inputs from different organizations and team participants to achieve project goals. Hence, ensuring effective communication is a vital requirement to effectively coordinate and accomplish the

T. E. Kwofie et al., *Effective Construction Project Delivery*,
https://doi.org/10.1007/978-3-030-49374-5_8

project and integrate the process. However, various factors account for the effectiveness of the communication among project teams in various procurements. In as much as these factors are considered critical to communication performance outcome, various studies have given significant attention in traditional procurement systems (Liu 2009; Xie 2002). Understanding the critical factors that impact on communication performance is seen as a veritable approach to improving communication in procurement typologies through the right actions. This chapter focuses on identifying the critical communication factors in non-traditional procurement. With this knowledge, communication in various typologies of non-traditional procurement could be improved leading to optimized benefits of these typologies.

8.2 Factors Influencing Communication Performance in Non-traditional Procurement

Communication in any type of procurement system is said to be very complex activity, and achieving its effectiveness is very difficult (Liu 2009; Liu and Shen 2008; Latham 1994; Pietroforte 1997; Xie et al. 2000; Dawood et al. 2002). Problems and challenges of effective communication across various typologies of procurement system in construction project delivery are significantly reported (see Guevara and Boyer 1981; Thomas et al. 1998; Xie et al. 2000; Xie 2002; Dawood et al. 2002; Kwofie et al. 2014, 2015, 2016, 2019a, b). Several studies have stated that various typologies of factors account for these communication problems in construction communication (see Marshall-Ponting and Aouad 2005; Pietroforte 1997; Thomas et al. 1998; Xie et al. 2000, 2010).

However, the persistent call for communication improvement in construction has cited technical and social actions as the fundamental theories towards improvement (Marshall-Ponting and Aouad 2005). The theoretical basis of the technical dimension focuses on tools and technology, whereas the social dimension sees soft skills, behaviours and concepts that can influence communication outcome. Hence, gaining an understanding of the factors that influence communication performance is deemed as a panacea towards improvement in construction communication (Liu 2009; Marshall-Ponting and Aouad 2005; Emmitt and Gorse 2003; Mead 1999; Kwofie et al. 2014, 2017; Keyton et al. 2013).

8.2.1 Technical Factors

Marshall-Ponting and Aouad (2005) stated technical tools and social action or human management approaches are the two main theories that account for communication improvement. From the technical dimension, the assertion is that technology and tools are the single factor that accounts for communication improvement in a given context (Marshall-Ponting and Aouad 2005). From this perspective, several studies have given credence to emerging technologies and tools such email, intranet, Internet, multimedia and virtual reality significantly changed and improved communication activities and process and performance in construction project delivery (El-Saboni et al. 2009; Huang et al. 2001; Mead 1999; Otter 2005; Shen 1992; Rojas and Songer 1999; Hardcastle 1990; Murray et al. 2000).

Mead (1999) and Otter accounted that the use of intranet and project website improved information sharing, interactions and communication management in construction project delivery. From the early 1980s to now, the use of technology and tools has improved considerably from helping basic manual tasks like booking, typing and number crunching to focused application on development of standalone computers for scheduling, estimation and design (Bouchlaghem et al. 2005; Fruchter 1996; Otter 2005; Sonnenwald 2006). The turn of 1990s was termed as the third phase which focused on IT developed as communication medium that is capable of establishing favourable supply chain relationship (Mead 1999; Liu 2009).

It is clearly noted that currently the communication environment in the construction project delivery process has changed significantly by adopting and applying information technology for improving communication through automatic information handling and expression, re-engineering the communication network, widening the prevailing communication network, handling bid data and information and bridging the virtuality gap (Kwofie et al. 2019b; Otter 2005; Huang et al. 2001; Fruchter 1996; Otter and Emmitt 2007). Pietroforte (1997) and Anumba and Evbuomwan (1999) acknowledged IT development in construction communication is significant in helping to fill the need for providing appropriate medium infrastructure for communication that facilitates multilateral and flexible communication among project teams. Shohet and Frydman (2003) and Austin et al. (1997) on the other hand provided empirical support to the fact that IT and other-related tools for the purposes of project team communication significantly improve communication. Hence, IT and other tools are deemed as very significant factors to communication performance in construction communication process and network improvement (Pietroforte 1997, 1999; Anumba and Evbuomwan 1999; Bouchlaghem et al. 2005; Otter 2005; Austin et al. 1997). Mead (1999) proved that the use of intranet increased communication effectiveness among the project team by allowing each timely access to vital project information by creating a virtual communication network that benefits communication performance. However, information and data compatibility problems were notable. However, the application of information modelling programmes and software such as IDEFO, IFO, STEP, ApePT, IFC and aecXML enabled the Internet overcome compatibility data format (Huang et al. 2001; Nitithamyong and Skibniewski 2006; Wix and See 1996).

Anumba and Evbuomwan (1999) accounted that other developments in IT for the construction communication and information sharing which entailed the development of visualization tools such as what you see is what I see (WYSIWIS) enabled the understanding of shared information in communication process in 3D and 4D format. Dawood et al. (2002) and Caldas and Soibelman (2003) all explored an automated communication system and prototype document classification to facilitate information management and data exchange among project teams and lend credence to the impact of ICT on communication performance in construction project delivery. Mead (1999) and Abduh and Skibniewski (2004) revealed that some information technology and tools such as groupwares adopted for communication offer improved communication performance by creating a collaborative environment for working, sharing information and interactions.

According to Castro-Lacouture and Skibniewski (2003) and Nitithamyong and Skibniewski (2006) explored some computer-aided tools and technology for collaborative work and found that such tools enhance electronic network and improved communication-supported operations and exchange of project-related information in highly distributed project teams and organization. Web-based project management technology has also been significantly used to facilitate information management and communication among project participants (Nitithamyong and Skibniewski 2006; Skibniewski and Abduh 2000; Huang et al. 2001; Abduh and Skibniewski 2004). It is worth noting that information technology and tools as a technical factors have significant impact on the communication effectiveness in construction project delivery given the array of evidence provided by the numerous studies. From this, it can be suggested that, indeed, ICT tools cannot be overlooked in communication improvement and collaborative work process.

8.2.2 Social/Human Management Factors

Arguments in favour of considering social actions or human organisational factors in communication performance towards improvement are of the theoretical position that, indeed, technical factors such as IT and tools aid communication, it must also be noted that these tools are operated by humans (Marshall-Pointing and Aouad 2005). Thus, technological solutions are not only tools supporting communication process because communication has a softer aspect which is also of significance. In the light of this, Marshall-Pointing and Aouad (2005) argued that a blend of both technical factors and social actions or human management factors must be pursued towards communication performance improvement. This position was shared by Otter (2005) and Emmitt and Gorse (2007).

Liu (2009) intimated that managing communication should also include soft aspects and people issues which encompass managing background of communicators, relationship between senders and receivers and the organizational and communication context. In construction project delivery, the way project teams are organized has become a significant social or human management factor that impact on communication (Liu 2009; Xie et al. 2010; Tran et al. 2017). Management approach and its impact on communication performance have been explored in various studies. Mead (1999) and Pietroforte (1997, 1999) indicated that ensuring the most appropriate contract arrangements sets up the communication lines, networks and information flow among the project teams. Pietroforte (1997) further established that organizational context and nature of the information are crucial factors in human management factors that should be considered in communication improvement strategies.

Shohet and Frydman (2003) found issues of formal relationships and contract types as major factors that impact on communication among project teams. Mead (1999) explored organizational typologies and found partnering arrangements and team building result in improved communication. Generally, there is an apparent admission among practitioners on the need to improve construction communication, especially in various procurement typologies. This is deemed not to be determinate but rather a continuous actions and strategies. In regard to ensuring communication improvement in construction, the general approach has been to explore the factor performance approach (see Liu 2009; Xie et al. 2010; Kwofie et al. 2015, 2016; Dawood et al. 2002; El-Saboni et al. 2009; Thomas et al. 1998). However, even though literature generally accepts the combination of technical factors and human management factor combination approach to communication improvement, the focus has overly been on technical specifically information technology (IT) and tools dimension.

Emmitt and Gorse (2003) bemoan the lack of attention to people issues in communication in construction and stated that the argument of adoption of information technology will transform work and communication was rather too optimistic. Other factors such as organizational relationships, social dynamics and context are equally important to be considered given their perceived impact and challenges it pose to communication (Pietroforte 1997; Shohet and Frydman 2003). Emmitt and Gorse (2003) contended that for significant gains to be made in improving communication in construction, more attention must be given to understanding the factors that influence communication performance. Few studies have explored the factors that impact on communication. The limited studies in these directions have approached it from different perspectives (see Guevara and Boyer 1981; Thomas et al. 1998, 1999; Shen 1992; Watkinson 1992; Wong et al. 2004; Xie 2002; Xie et al. 2010).

Wong et al. (2004) and Xie (2002) explored project attributes, organizational context, individual characteristics, communication variables and management strategies as factors that could improve communication in project delivery. Under communication variables, Guevara and Boyer (1981) identified message characteristics, feedback, completeness of the information, distortion, gatekeeping, information processing, centrality, source credibility, modality choice, speed, directionality, accuracy, activity level and network alignment as key factors. In the area of individual attributes, Guevara and Boyer (1981) further revealed status authority, influence, expectations, mobility aspiration, satisfaction with communication, trust and interactions as key factors. At the organizational factors level, studies explain hierarchy size, structure, technology workflow, nature of formal relationships, trust, mutual respect, rules and norms of organization and performance criteria are significant factors (Guevara and Boyer 1981; Wong et al. 2004; Tran et al. 2017).

Wong et al. (2004) further accounted industry culture, organizational culture, workers attitude and values of communicators as factors that impact on communication of safety information. Classification of social or human factors impacting communication performance as individual attributes, communication variables and organizational characteristics is consistent with various studies (see Guevara and Boyer 1981; Thomas et al. 1998, 1999; Mead 1999; Xie 2002; Wong et al. 2004; Dawood et al. 2002).

8.3 Nature of Influence of the Factors to Communication Performance

Exploring nature of factors that impact on communication is deemed as a consistent approach towards communication improvement. This has been adopted in many studies (Thomas et al. 1998; Xie 2002; Dawood et al. 2002; Liu 2009; Kwofie 2015; Kwofie et al. 2015, 2016; Shen 1992; Watkinson 1992). By using a structural equation modelling (SEM), the study explored the impact of individual attributes, communication variables and organizational characteristics on communication performance in partnering, project alliance, integrated project delivery and PPP arrangements in project delivery.

From a survey of 241 responses, the results of the model fit of the structural equation modelling revealed a good fit to the data with the Chi square (χ^2) = 2913.48, $p < 0.000$, CFI = 0.984, GFI = 0.927 and RMSEA = 0.002. From the works of Iacobucci (2010) and Kline (2010), a robust GFI and CFI indexes between 0.950 and 1.00 (upper limit) are said to be a good fit. Likewise, a very good fit model will have the RMSEA value up to 0.05. The value of 0.002 for RMSEA is an indication of a good fitting model (Iacobucci 2010; Kline 2010; Bentler 2005). On the score of the nature of influence of the individual attributes, communication variables and the organizational characteristics, the results indicated that the coefficient of determination (R^2) was 0.683. This suggests that the three factors (individual attributes, communication variables and organizational characteristics) contribute to 68.3% of the variance in communication performance in non-traditional procurements systems (partnering, project alliancing, IPD and PPP). The path coefficient loadings (effects) of the individual attributes, communication variables, organizational characteristics are the composite (total) effect of the factors (R^2) on the communication performance factor which is often interpreted as the predictive accuracy of the model evaluated (Hair et al. 2014; Iacobucci 2010; Kline 2010).

Hair et al. (2014) revealed that the statistic effect ranges from 0.00 to 1.00 with 1.00 suggesting an absolute predictive accuracy. Thus, R^2 values of 0.75, 0.50 and 0.25 describe substantial, moderate or weak levels of predictive accuracy, respectively (Hair et al. 2014; Iacobucci 2010). In the insight provided by Iacobucci (2010) and Frank and Hennig-Thurau (2008), an R^2 value less than 0.100 is counted as an insignificant effect on the outcome variable. Hence, the factor loading indicates that individual attributes contribute 19.8% (0.198), communication variables contribute 53.7% (0.537), whereas that of organizational characteristics was 58.1% (0.581) of the total 68.3.2% variance in the communication performance outcome.

From a critical assessment of these results, it can be said that communication variables contribute (53.7%) and organizational characteristics (58.1%) had the strongest impact (contribution) to communication performance in non-traditional procurements adopted on the projects delivered. This presents an empirical support to the contribution of social actions or organizational management factors to communication performance. Partly, it can be stated that this findings affirm the revelation by Xie et al. (2010) and Tran et al. (2017) that nature of relationships, quality of networks and trust levels among others impact on the level of communication performance and information sharing in partnering and alliance contracting. Emmitt and Gorse (2007) and Gorse and Emmitt (2007) provided an insight on communication behaviours adopted by project team participants. From this, it can be said the aspect of the contribution of individual attributes seems to support Emmitt and Gorse (2007) and thus can be said that communication behaviours among project teams is a valid factor that contribute to communication performance in non-traditional procurement systems.

8.4 Systematic Communication Strategy of the Factors for Communication Improvement

Christensen (2014) provided both theoretical and practical insight to the fact that communication can be used as strategic tool and process to influence change process, maintain team dynamics and improve relationship in teamwork and task function. Many construction organization and project teams today in project delivery process are continuously confronted with the need for adopting strategic communication that can improve relationship, build trust, engender mutual respect and deal with change without adverse effect. An obvious major challenge most project organizations and project teams face is to cope with contextual communication challenges that allow them develop or adopt both technology, tools and human organizational factors such culture, leadership that can create the acceptable environment capable of ensuring systematic and strategic communication towards communication performance improvement (Dawood et al. 2002; El-Saboni et al. 2009).

One of the approaches to a seemingly successful improvement in communication improvement in construction project delivery through the adoption and establishment of effective communication and information strategies among teams and within the project organizations while dealing with issues of trust, relationship improvement, conflict and change process (Salleh 2008; Henderson 2008). Bordia et al. (2004) and Barrett (2002) posited that communication is well recognized as an effective tool and is instrumental to team effectiveness and survival of organizations as well as team bonding. Bordia et al. (2004) further iteratively provided evidence to the fact that developing continuous communication audit can serve as a strategic system that can create the foundation for diagnosing the quality of the communication practice.

Communication auditing entails the assessment of current communication strategies and practices to determine what actions are required for improvement.

This line of thought has led to the development of various instruments for measuring and evaluating team and organizational communication, but these existing measures are diverse and differ in their aims regarding communication focus. Incidence of mistrust, distortions, gatekeeping, centrality, source credibility, hierarchy size, inconsistent rules and norms as well as inaccuracies in communication are all significant factors that affect team functions, cohesion, organizational effectiveness that can hinder progress in project delivery (Guevara and Boyer 1981; Wong et al. 2004). They can also pose enormous threats to interpersonal relationships, impact on internal processes and systems, and on social functions (Pardo del Val and Martinez Fuentes 2003; Van Vuuren et al. 2007; Goldhaber 1999).

Waddell et al. (2011) and Van Vuuren et al. (2007) contended that strategic communication in openness and honesty can significantly reduce anxiety of the threats of mistrust. Waddell et al. (2011) also intimated that effective strategy to eliminate the impact of various communication factors is to involve all participants and organization in the planning and process of communication as well as allowing easy and uncontrolled access to shared information. This approach has a high propensity to increase confidence of the teams and organizations as well as their interest and motivate them to freely share and communicate information for the purpose of the project. According to Goldhaber (1999), the concept of strategic team or organizational communication is significantly marked by diversity and this is seen in the many definitions which reflect a wide range of approaches and perspectives. From the analysis of the content perspective, Goldhaber (1999) exposed three common features of strategic team or organizational communication, where enumerated as (a) takes place within complex open systems (influenced by the environment and influences it as well); (b) entails all features of a message (i.e. flow, purpose, direction, media, centrality and directionality); and (c) involves people, including their attitudes, feelings, relationships and skills.

From Downs et al. (1994) and Keyton et al. (2013), there is an obvious fact that strategic team and organization communication demands a broad perspective. In the work of Downs et al. (1994) which was further tested and validated by Christensen (2014), social contact, central leadership, information, emotive influence and barriers to improvement are identified as the four main categories of strategic communication process in team and organizational work environment of the factors for communication improvement. The social contact category of strategic communication identifies several elements as crucial to well-functioning strategic communication systems in teamwork and organization functioning in delivery process. In this category Christensen (2014) validated contact between coworkers and leadership, contact among coworkers, avoiding disruptive relations between colleagues and leaders, distinguishing between contact with central leadership and contact with the closest leader as significant themes that can create the social cohesive environment for teamwork, task function and organizational effectiveness.

The central leadership strategic category posits that communication and information processes could be experienced quite differently based on who initiates the rationale and strategy behind a planned process. Against this, the central leadership follows the approach of leaders being central to the communication process and use their persuasive, affective, innovative attributes positively to influence teams and organizations (Goldhaber 1999). By this, leaders are able to affect the attitudes, feelings, relationships and skills of team members and organisations (Tukiainen 2001; Keyton et al. 2013). In the information strategic category, themes of information content, quality, timing and structure are mentioned as the tenets for enhancing communication (Johnson 1992; Tukiainen 2001). According to Rafferty and Jimmieson (2010), the information climate regarding teamwork and organizational function and processes has an influence on the quality of work life, work overload and distress.

The fourth category considers the role of influence as a strategic communication tool (Tukiainen 2001; Downs et al. 1994). This category sees participation as a perceived influential tool that might function as a predictor of acceptance of decision, increase commitment and empowerment to team members (Gagné et al. 2000; Wanberg and Banas 2000). The fifth category was concerned with barriers to the improvement of communication during organizational and teamwork process (Koontz 2001; Bruckman 2008). Koontz (2001) summarized barriers to communication with the notion that communication problems are often symptoms of problems that are more deeply rooted. Hence, by adopting transparent approach to communication and increasing technology penetration, most barriers could be eliminated (Waddell et al. 2007; Bruckman 2008; Erwin and Garman 2009; Ford and Ford 2009).

8.5 Summary

Here in this chapter, a good insight into communication factors that impact on communication performance has been elucidated. The results further show significant contribution of these individual attributes, communication variables and organizational characteristic factors to communication performance outcome. Likewise, a systematic understanding of the nature of the influence has also been give. Five-tier strategic communication categories which can ameliorate the negative impact of these communication factors have also been illuminated.

References

Abduh, M., & Skibniewski, M. J. (2004). Electronic networking technology in construction. *Journal of Construction Research, 5*(1), 17–42.

Anumba, C. J., & Evbuomwan, N. F. O. (1999). A taxonomy for communication facets in concurrent life-cycle design and construction. *Computer-Aided Civil and Infrastructure Engineering, 14*(1), 37–44.

Austin, S. A., Baldwin, A. N., & Newton, A. (1997). Manipulating the flow of design information to improve the programming of building design. *Construction Management and Economics, 12*, 445–455.

Barrett, D. J. (2002). Change communication: Using strategic employee communication to facilitate major change. *Corporate Communications: An International Journal, 7*(4), 219–231.

Bentler, P. M. (2005). *EQS 6 structural equation program manual*. Encino, CA: Multivariate Software.

Bordia, P., Hobman, E., Jones, E., Gallois, C., & Callan, V. J. (2004). Uncertainty during organizational change: Types, consequences, and management strategies. *Journal of Business and Psychology, 18*(4), 507–532.

Bouchlaghem, D., Shang, H., Anumba, C. J., Cen, M., Miles, J., & Taylor, M. (2005). ICT-enabled collaborative working environment for concurrent conceptual design. *Architectural Engineering and Design Management, 1*(4), 261–280.

Bruckman, J. C. (2008). Overcoming resistance to change: Causal factors, interventions, and critical values. *The Psychologist-Manager Journal, 11*, 211–219.

Caldas, C., & Soibelman, L. (2003). Automating hierarchical documents classification for construction management information systems. *Automation in Construction, 12*(4), 395–406.

Castro-Lacouture, D., & Skibniewski, M. J. (2003). Applicability of e-Work models for the automation of construction materials management systems. *Production Planning and Control, 14*(8), 789–797.

Christensen, M. (2014). Communication as a strategic tool in change processes. *International Journal of Business Communication, 51*(4), 359–385.

Dawood, N., Akinsola, A., & Hobbs, B. (2002). Development of automated communication of system for managing site information using internet technology. *Automation in Construction, 11*(3), 552–572.

Downs, C. W., DeWine, S., & Greenbaum, H. H. (1994). Measures of organizational communication. In R. B. Rubin, P. Palmgreen, & H.-E. Sypher, et al. (Eds.), *Communication research measures: A sourcebook* (pp. 57–78). New York, NY: Guilford Press.

El-Saboni, M., Aouad, G., & Sabouni, A. (2009). Electronic communication systems effects on the success of construction projects in United Arab Emirates. *Advanced Engineering Informatics, 23*(1), 130–138.

Emmitt, S., & Gorse, C. A. (2003). *Construction communication*. Oxford: Blackwell Publishing.

Emmitt, S., & Gorse, C. A. (2007). *Communication in construction teams*. London: Taylor & Francis.

Erwin, D., & Garman, A. (2009). Resistance to organizational change: Linking research and practice. *Leadership & Organization Development Journal, 31*(1), 39–56.

Ford, J., & Ford, L. (2009). Resistance to change: A reexamination and extension. *Research in Organizational Change and Management, 17*, 211–239.

Frank, B., & Hennig-Thurau, T. (2008). Identifying hidden structures in marketing's structural models through universal structure modeling: An explorative Bayesian neural network complement to LISREL and PLS. *Marketing—Journal of Research and Management, 4*(2), 47–66.

Fruchter, R. (1996). Interdisciplinary communication medium in support of synchronous and asynchronous collaborative design. In B. Kumar & A. Retik (Eds.), *Information representation and delivery in civil and structural engineering design* (pp. 21–28). Edinburgh, UK: Civil Comp Press.

Hair, J. F., Sarstedt, M., Hopkins, L., & Kuppelwieser, V. J. (2014). Partial least squares structural equation modeling (PLS-SEM): An emerging tool in business research. *European Business Review, 26*(2), 106–121.

Huang, G. Q., Yee, W. Y., & Mak, K. L. (2001). Development of a web-based system for engineering change management. *Robotics and Computer Integrated Manufacturing, 17*(3), 255–267.

Gagné, M., Koestner, R., & Zuckerman, M. (2000). Facilitating acceptance of organizational change: The importance of self-determination. *Journal of Applied Social Psychology, 30*(9), 1843–1852.

Goldhaber, G. M. (1999). Organizational communication in 1976: Present domain and future directions. In P. Salem (Ed.), *Organizational communication and change* (pp. 4–30). Cresskill, NJ: Hampton Press.

Gorse, C. A., & Emmitt, S. (2007). Communication behaviour during management and design team meetings: A comparison of group interaction. *Construction Management and Economics, 25,* 1197–1213.

Guevara, J. M., & Boyer, L. T. (1981). Communication problems within construction. *Journal of Construction Engineering, American Society of Civil Engineers, 107*(4), 552–557.

Hardcastle, C. (1990). Information transfer between design and construction. In *Proceedings of CIB 90—Building Economics and Construction Management* (Vol. 2, pp. 12–21), March 14–21, Sydney.

Henderson, L. S. (2008). The impact of project managers' communication competencies: Validation and extension of a research model for virtuality, satisfaction, and productivity on project teams. *Project Management Journal, 39*, 48–59.

Iacobucci, D. (2010). Structural equations modeling: Fit indices, sample size, and advanced topics. *Journal of Consumer Psychology, 20,* 90–98.

Johnson, J. D. (1992). Approaches to organizational communication structure. *Journal of Business Research, 25*(2), 99–113.

Keyton, J., Caputo, J. M., Ford, E. A., Fu, R., Leibowitz, S. A., Liu, T., et al. (2013). Investigating verbal workplace communication behaviors. *Journal of Business Communication, 50*(2), 152–169.

Kline, R. B. (2010). *Principles and practice of structural equation modeling* (3rd ed.). New York: Guilford Press.

Koontz, H. (2001). *Management: A global perspective* (10th ed.). New York, NY: McGraw-Hill.

Kwofie, E. (2015). *Contribution of unique features of mass housing projects to project team communication performance* (Unpublished Ph.D. thesis). Department of Building Technology, Kwame Nkrumah University of Science and Technology, Kumasi.

Kwofie, T. E., Adinyira, E., & Fugar, F. (2017). Theoretical and practical implications for engendering project team communication effectiveness in mass housing project delivery in Ghana. *Journal of Engineering, Design and Technology, 15*(6), 826–844.

Kwofie, T. E., Aigbavboa, C. O., & Machethe, S. O. (2019a). Nature of communication performance in non-traditional procurements in South Africa. *Engineering, Construction and Architectural Management, 26*(10), 2264–2288.

Kwofie, T. E., Aigbavboa, C. O., & Thwala, W. D. (2019b). Communication performance challenges in PPP projects: Cases of Ghana and South Africa. *Built Environment Project and Asset Management, 9*(5), 628–641.

Kwofie, T. E., Fugar, F., & Adinyira, E. (2015). Contribution of multiple construction site management features to project team communication effectiveness: The case of mass housing projects. *Engineering Project Organization Journal, 5,* 180–193.

Kwofie, T. E., Fugar, F., & Adinyira, E. (2016). Modelling the effect of housing design unit contract packaging on mass housing project team communication performance. *Journal of Construction in Developing Countries, 21*(1), 35–50.

Kwofie, T. E., Fugar, F. D. K., Adinyira, E., & Ahadzie, D. K. (2014). A conceptual framework for evaluating communication performance among mass housing project team. In O. Ejohwomu & O. Oshodi (Eds.), *Proceedings of 3rd International Conference on Infrastructure Development in Africa* (pp. 6–21), March 17–19, 2014, Abeokuta, Nigeria.

Latham, S. M. (1994). *Constructing the team.* Joint Government/Industry Review of Procurement and Contractual Arrangements in the UK Construction Industry. Final Report, HMSO, London.

Liu, Y. (2009). *Critical factors for managing project team communication at the construction stage* (Ph.D. thesis). Submitted to the Polytechnic University of Hong Kong.

Liu, Y., & Shen, L. (2008). Factors influencing project team communication and its management. In *Proceedings, CRIOCM 2008 International Research Symposium on Advancement of Construction Management and Real Estate* (pp. 219–223), 31 October–3 November 2008, Beijing, China.

Marshall-Ponting, A. J., & Aouad, G. (2005). An nD modelling approach to improve communication process for construction. *Automation in Construction, 14*(3), 311–321.

Mead, S. T. (1999). *Communication effectiveness in intranet based construction projects* (Unpublished Ph.D. thesis). Department of Civil and Building Engineering, Loughborough, University, Loughborough, UK.

Murray, M. D., Tookey, J. E., Langford, D. A., & Hardcastle, C. (2000). Project communication variables: A comparative study of US and UK construction industry perceptions. In *Proceedings of ARCOM Conference* (pp. 813–822), Glasgow.

Nitithamyong, P., & Skibniewski, M. (2006). Success/failure factors and performance measures of web-based construction project management systems: Professionals' viewpoint. *Journal of Construction Management and Engineering, 132*(1), 80–87.

Otter, A. F. H. J. (2005). *Design team communication using a project website*, Ph.D. thesis, Bouwstenen 98, Technische Universiteit Eindhoven, Eindhoven.

Otter, A., & Emmitt, S. (2007). Exploring effectiveness of team communication: Balancing synchronous and asynchronous communication in design teams. *Engineering, Construction and Architectural Management, 14*(5), 408–419.

Pardo del Val, M., & Martinez Fuentes, C. (2003). Resistance to change: A literature review and empirical study. *Management Decision, 41*(2), 148–155.

Pietroforte, R. (1997). Communication and governance in the building process. *Construction Management and Economics, 15*(1), 71–82.

Pietroforte, R. (1999). *Communication and information in the building delivery process* (Ph.D. thesis). Massachusetts Institute of Technology, Cambridge, MA, USA.

Rafferty, A. E., & Jimmieson, N. L. (2010). Team change climate: A group-level analysis of the relationships among change information and change participation, role stressors, and well-being. *European Journal of Work and Organizational Psychology, 19*(5), 551–586.

Rojas, E. M., & Songer, A. D. (1999). Web-centric systems: A new paradigm for collaboration engineering. *Journal of Management and Engineering, 15*(1), 39–45.

Salleh, L. M. (2008). Communication competence: A Malaysian perspective. *Pacific and Asian Communication Association, 11*(3), 303–312.

Shen, L. Y. (1992). *Information management in construction companies*. Hong Kong: Department of Building and Surveying, Hong Kong Polytechnic University.

Shohet, I. M., & Frydman, S. (2003). Communication patterns in construction at construction manager level. *Journal of Construction Engineering & Management, 129*(5), 570–577.

Skibniewski, M. J., & Abduh, M. (2000). Web-based project management for construction: Search for utility assessment tools. In *Proceedings of INCITE* (pp. 56–77), 2000, Hong Kong Polytechnic University, Hong Kong.

Sonnenwald, D. H. (2006). Challenges in sharing information effectively: Examples from command and control. *Information Research—An International Electronic Journal, 11*(4), 251.

Thomas, S. R., Tucker, R., & Kelly, W. (1998). Critical communications variables. *Journal of Construction Engineering Management, 129*(1), 58–66.

Thomas, S. R., Tucker, R., & Kelly, W. (1999). Compass: An assessment tool for improving project team communications. *Project Management Journal, 30*(4), 15–23.

Tran, D. Q., Nguyen, L. D., & Faught, A. (2017). Examination of communication processes in design-build project delivery in building construction. *Engineering, Construction and Architectural Management, 24*(6), 1319–1336.

Tukiainen, T. (2001). An agenda model of organizational communication. *Corporate Communications: An International Journal, 6*(1), 47–52.

Van Vuuren, M., de Jong, M., & Seydel, E. (2007). Direct and indirect effects of supervisor communication on organizational commitment. *Corporate Communications: An International Journal, 12*(2), 116–128.

Waddell, D., Cummings, T., & Worley, C. (2007). *Organization development and change*. Pyrmont, NSW, Australia: Nelson ITP.

Waddell, D., Cummings, T., & Worley, C. (2011). *Organizational change: Development & transformation*. South Melbourne, Australia: Cengage Learning.

Wanberg, C. R., & Banas, J. T. (2000). Predictors and outcomes to changes in a reorganizing workplace. *Journal of Applied Psychology, 85*(1), 132–142.

Watkinson, M. (1992). Procurement alternatives. *Faculty of Building Journal*, 4–7.

Wix, J., & See, R. (1996). *An introduction to the international alliance for interoperability and the industrial foundation classes*. VA, USA: Oakton.

Wong, F., Chan, A., Fox, P., Tse, P., & Ly, E. (2004). *Identification of critical factors affecting the communication of safety-related information between main contractors and sub-contractors in Hong Kong*. Report.

Xie, C., Wu, D., Luo, J., & Hu, X. (2010). A case study of multi-team communications in construction design under supply chain partnering. *Supply Chain Management: An International Journal, 15*(5), 363–370.

Xie, X. (2002). *Communications in construction design* (Ph.D. thesis). Loughborough University, Loughborough.

Xie, X., Thorpe, A., & Baldwin, A. N. (2000). A survey of communication issues in construction design. In A. Akintoye (Ed.), *16th Annual ARCOM Conference* (Vol. 2, pp. 771–780), September 6–8, 2000, Glasgow Caledonian University. Association of Researchers in Construction Management.

Part III
Improving Construction Communication

Chapter 9
Communication Dynamics in Non-traditional Procurement Communication

Abstract The conceptual issues underlining teamwork and group dynamics are said to be very complex, and its implications for communication, team effectiveness and management are not extensively understood, especially in non-traditional procurement typologies. The case of project team and groups exhibiting complex social and task-related communication network is well acknowledged but unfortunately not rigorously interrogated. Group work dynamics, incidence of team culture and nature of social networks have increasingly become key factors in construction project design management, effectiveness of the process and information sharing and team interactions. They also offer significant implications for collaboration, trust and conflict-related issues which are key tenets of non-traditional procurement typologies. This chapter offers theoretical and conceptual insights into the dynamics of teamwork and group work and explores the complexities of their social networks and its impact on communication effectiveness and interactions in project delivery using social network analysis (SNA). Through this, the communication patterns, the communication network can provide the level of description and performance inherent in the team and group dynamics and culture. This analysis is able to describe network relations, identify prominent patterns, trace information flow, and measure the strength of network ties in complex organisational communication networks among project teams and groups. This will help project teams to focus their attention on the necessary aspects of multi-team and group communications inherent from the team dynamic and culture in various non-traditional procurement models towards improving collaboration, cooperation and integration in teamwork.

Keywords Teamwork dynamics · Social network analysis · Team effectiveness · Communication patterns · Communication networks

9.1 Introduction

Construction project design and delivery has become an increasingly complex activity, and there is an increasing awareness of the need for better design management in the construction industry, especially across various procurement typologies. Group work dynamics, incidence of team culture and nature of social networks have

T. E. Kwofie et al., *Effective Construction Project Delivery*,
https://doi.org/10.1007/978-3-030-49374-5_9

increasingly become key factors in construction project design management and effectiveness of the process. The factors of team dynamics, incidence of team culture and nature of social networks provides both theoretical and practical perspective to view communication in teams from different dimensions which are often ignored. The potential benefits of integrating group work dynamics, team culture and nature of social networks into early communication planning and strategies are widely recognized in construction communication literature. This chapter explores the tenets of group work dynamics, incidence of team culture and social network in construction communication. It is argued that understanding these factors in construction communication can help the industry to develop bespoke tools, competency behaviours and models that can optimize communication performance outcome in construction project delivery.

9.2 Teamwork and Group Work Dynamics in Non-traditional Project Teams

As emerging project and construction environments have become even more competitive and forming complex networks, the work and team dynamics presents project teams to make significant efforts to operate internal and external linkages within and across team and organizational boundaries (Lee and Sawang 2016; Yang et al. 2011; Chou and Yang 2012; Bond-Barnard et al. 2013). The dynamics exhibited by work teams or groups are inherent from their creation or composition, development, operation, management nature and flow of work or task (Kozlowski and Bell 2013). Kozlowski and Bell (2013) indicated that increasing global competition, consolidation and innovation create pressures that are influencing the emergence of work teams as the core building blocks of organizations and task functions which is driving a need for a more rapid, flexible and adaptive responses to teamwork and work groups. This is because these dynamics needs to be understood so as to develop necessary diverse skills, expertise and experience for teams.

The dynamics of teamwork and group work in non-traditional procurement systems necessitate more rapid, flexible and adaptive responses to task functions, organizational process and operational functions in project delivery. This creates an urge for creativity, invention and innovation that allows for coping with globalized operations through expansion, mergers and acquisitions, and joint ventures placing increased importance on cross-cultural, mixed culture and boundary-spanning teams (Kozlowski and Bell 2008, 2013; Lee and Sawang 2016; Kozlowski et al. 2011; Marrone 2010; Crawford and Lepine 2013). In construction project delivery, the ability of teams and organizations to exploit the potential of work group and team, it is important for the participants to have an awareness of the dynamics that are at play in teamwork. Some of these dynamics tend to affect the production of information and forms of interaction in groups and teams (Gorse and Sanderson 2007; Gorse et al. 2006a; Sutton 2006).

Construction project delivery and task functions have dominantly become team based in every aspect. Indeed it can unequivocally be accepted that groups and team are fundamental to the construction project delivery across almost all procurement systems and management models (Gorse and Sanderson 2007; Gorse et al. 2006b; Crawford and Lepine 2013). In spite of the significance of the team approach to project delivery and task function, it is clear to note that generally little thought is given to how teams function in a group, the types of behaviour they exhibit and the teeming dynamics that affect work and tasks. From the perspective of team creation and creation or composition, development, operation, management nature, flow of work or task, group attachment, project teams and work groups exhibit varied dynamics (Lee and Sawang 2016; Lee 2005; Lee and Ling 2007).

One area which has become an essential dynamics of teams and groups is team boundaries (Bond-Barnard et al. 2013; Crawford and Lepine 2013). Bond-Barnard et al. (2013) professed that in achieving project efficiency project teams and work groups depend increasingly on communications and collaborations across team boundaries team members operate external linkages within or across organizational boundaries. When the work and task environment becomes increasingly competitive, team members venture beyond team boundaries to seize innovation opportunities to perform (Crawford and Lepine 2013). However, some participants of the team boundary-span less extensively, which isolate themselves and their project teams from external environments (Lee and Sawang 2016; Lee 2005). This brings into perspective externally focused behaviour of teams and groups. Teams and work groups deemed boundary spanners in tasks and operational processes need to deal with interpersonal relationships and project environments inside and outside their teams (Friedman and Podolny 1992; Puurtinen et al. 2015). This impacts on their relational orientation, their behaviours, interactions and attachment to the team. In then considered views of Lee (2005) and Lee and Ling (2007) team, attachment is an individual-level construct based on an individual's perception of his or her relationship to the specific team, and this provides a psychological foundation of team boundary management.

Lee and Sawang (2016) iteratively provided insights that point to the fact that some project team participants excel while others tend to derail tasks and teams in external activities that span across their team boundary through team or group attachment. It is suggested that attachment to teams and groups are inherent from individuals' psychological ties to their groups as a whole, rather than to another person (Lee 2005; Lee and Ling 2007; Smith et al. 1999). These attachments tend to influence relational dimensions, information sharing and communication, trust and interactions (Marmarosh and Tasca 2013; Marrone 2010; Rom and Mikulincer 2003). Even though attachment to teams and groups appear to be an individual-level construct, there is enough empirical demonstration of the fact that group members have different levels of attachment to the team or group (Rom and Mikulincer 2003).

Lee and Ling (2007) and Rom and Mikulincer (2003) deduced that team and group attachment has significant influence on individuals' cognition, emotion and behaviour in task-related and social groups which are a true attribute of team dynamics. In Smith et al. (1999), it is construed that attachment to teams and groups can have group attachment anxiety and group attachment avoidance as the two notable distinct dimensions. These dimensions tend to influence the performance in relations to task-based and social-based interactions (Emmitt and Gorse 2007). Team and group attachment anxiety induces inconsistent and unpredictable reactions from team members to individuals' fear-motivated support-seeking behaviours, whereas group attachment avoidance is a reflection of the extent to which an individual distrusts group members' goodwill and strives to maintain autonomy and emotional distance from them (Smith et al. 1999). High group attachment anxiety leads to low self-confidence and associates their team with inconsistent support and respect (Lee and Ling 2007). Lee and Ling (2007) did not discount the effects of boundary spanning attachment and insisted that high group attachment avoidance often distrust their teams which cause many individual team or group members to remain self-reliant and emotionally distant.

Others have also espoused other forms of team and group dynamics that significantly impact on team functions and interactions (see Belbin 1981, 1993, 2000; Bell and Kozlowski 2011). It is reckon that project teams and groups in various alliances, partnerships, ventures and coalition continuously be developed to encourage greater awareness of the team and group dynamics and how members can be effectively managed to optimize benefits and limit their weakness (Emmitt and Gorse 2007; Gorse et al. 2006a; Gorse and Sanderson 2007). In Gorse and Sanderson (2007), it was observed that sometimes people in team and group work are more comfortable and enjoy listening and communicating in group settings and evoke a satisfying experience if the group is effective especially in construction management. On the contrary, there are enough evidence that suggests that teamwork and task function can induce fear, generate boredom and make people dissatisfied and annoyed which can be a worrying dynamic of the team (Emmitt and Gorse 2007; Gorse and Sanderson 2007). The expression that non-traditional procurement systems exhibit varied group and interpersonal dynamics is an important part of understanding teamwork. Dainty et al. (2006) provided insights that support that when members become overly focused on minor issues by taking entrenched position, there can be an incidence of a general dip in team or group performance.

Bell (2001) and Gorse's (2002) from a critique of teamwork theoretically affirmed that interactions in team and group works become unevenly skewed with some members playing a passive and minimal role, being reluctant to participate, while others dominate interaction. This dynamics leads to communication breakdown, poor team effectiveness and strained relational orientations (Kwofie et al. 2015; Emmitt and Gorse 2003; Brown 2000). Some studies have suggested that to stay clear of these dynamics that weaken team and group work, it is important for team and group participants to continuously engage and add their information to the group's pool of knowledge and also develop their group interaction skills (Gorse and Sanderson 2007; Brown 2000).

Some key areas of teamwork dynamics that has gained continuous credence and attention has been on virtual teams, leadership, decision-making, team diversity, multi-team systems, team learning and macrocognition (Kozlowski and Ilgen 2006; Devine et al. 1999; Gully 2000; Gully et al. 2002; Guzzo and Dickson 1996). In some opinion, construction project teams formed by collaborative working model and integrated tend to be more stable, have clear leadership, prompt on decisions and embrace team learning culture (Chen et al. 2009, 2011; Kozlowski and Ilgen 2006; Gully et al. 2002). This is inherent from the non-adversarial nature of the teams which promote good and healthy interactions in a non-fierce competitive environment. Even though some scholars distinguish work teams and work groups (e.g. Gully et al. 2002; Katzenbach and Smith 1993), this work makes no such distinction and typically of construction project management approach to team and group work assume them as same with similar traits and thus used interchangeably. The fundamental dynamic is that project teams and groups evolve in a variety of types and sizes, cutting across different contexts, functions, internal processes and external linkages which impact on its effectiveness, information sharing and interactions (Kozlowski and Bell 2003; Arrow et al. 2000).

9.3 Incidence of Team Culture and Communication in Non-traditional Teams

The effect of culture on team interactions, information sharing and communication in project teams and organisations has increasingly become important and gained attention due to its significance to influencing the establishment of a sound alliancing approach to projects (Phua and Rowlinson 2003). Project alliancing, partnering, coalition, joint ventures and strategic supply chain involve numerous participants from different political, legal, economic and cultural backgrounds. With culture being one of the major issues affecting the management of construction projects, it is well acknowledged that its antecedents cannot be overlooked if managerial effectiveness is to be achieved in construction project delivery. There are valid foundation that offers empirical support to the impact of cultural issues on various contractual arrangements, conflict causation and nature of interaction among teams across numerous procurement systems in the delivery of construction projects (Vorster 1993; Diekmann and Girard 1995; Cheung 1998; Hofstede 1997).

Generally, it is accepted that procurement arrangements such as alliancing, coalition, joint ventures, partnering and strategic supply chains involves people from different backgrounds and cultural orientations that can be national or international (Xue et al. 2010; Chan and Tse 2003; Sambasivan and Yen 2010). Ensuring collaboration and integration of project delivery in such procurement systems demands greater cultural understanding and sensitivity (Sambasivan and Yen 2010). This can lead to several human-related problems such as language, communication and the lack of understanding of cultural differences (Chan and Tse 2003). Cultural issues in non-traditional procurement systems often includes systems of values, style, attitudes, orientation and norms in doing things, codes of conduct and expectations that guide behaviour (Loosemore 1999; Kumaraswamy and Yogeswaran 1998).

In project teams, their culture reflects social system through which they interact and communicate with each other and maintain order (Chan and Tse 2003; Sambasivan and Yen 2010). Considering the diversity in alliancing, partnering, coalition, joint ventures and strategic supply chains, it can be asserted that all parties involved in such procurement arrangements involved must be cross-culturally competent (Tso 1999; Trompenaars and Williams 1999). To this end, Trompenaars and Williams (1999) claimed that universalism versus particularism, individualism versus communitarianism, specific versus diffuse, neutrality versus affectivity, inner directed versus outer directed, achieved versus ascribed status and sequential versus synchronic time are the seven tenets that must be managed to ensure cultural competent teams in project delivery arrangements. These are complex psychological and behavioural dimensions of cultural issues in project teams that are often overlook.

Cultural issues in non-traditional procurement systems contribute to conflicts among parties and increase significantly the difficulties in the management of project and process (Fellow and Hancock 1994). It must be noted that it is not always that cultural issues are negative. In fact, project teams can appropriate certain attributes of culture to increase the effectiveness of the teams and performance (Tso 1999; Chan and Tse 2003). At the heart of cultural issues in construction project delivery lies a deeper layer of embedded behavioural traits that epitomizes culture that underlines their values and systems which eventually influence the communication and information sharing in teams (Chan and Tse 2003).

Culture is also conceptualized as knowledge, behaviour and value and evolves over time (Trompenaars 1993). In project teams, the culture of teamwork improves as the team develops synergy overtime and shape their collective activities. Communication and information sharing also suffer from effects of organizational culture (Xue et al. 2010; Wood et al. 2001). Tierney (1988) revealed that organisational culture is an interconnected web of relationships within or/and across other organizations. From this dimension, the work of Handy (1985) and Hofstede (1991) highlighted power culture, role culture, task culture and person culture as the four primary forms of organisational culture. The power culture is often configured as a web with the primary power at the centre and is frequently found in entrepreneurs (Handy 1985; Hofstede 1991). Role culture on the other hand normally relates to the functions and professional roles serving the structural pillars to support the overarching top management. In Handy (Handy 1985; Hofstede 1991), the case of task culture refers to the basic unit of the net underlining the job or project orientation whereas person culture forms the basic unit in which people interact and cluster relatively freely.

Wood et al. (2001) provided a theoretical position on team culture and expressed that cultures vary in their underlying values and attitudes. Rowlinson (2001) has given an in-depth exposition of this. From the core function of communication, information sharing and interactions in project delivery, it is contended that without understanding of the cultural orientations of teams and organisation there can be no cultural effectiveness. Hence, project teams and organizations must first identify with the values, social systems, attitudes, language and customs if teams can be culturally effective to perform in project delivery.

Recent times have witnessed growth in the number of strategic alliances, partnering, coalition and joint venture primarily due to the effects of shrinking world markets and increased competition (Pett and Dibrell 2001). Even though this phenomenon is deemed as a positive outcome, through active engagement in these strategic alliances, their inter-firm linkages have been frequently experienced problems of instability, poor performance and termination (Parkhe 1993). This has led to an exploration of the factors that lead to success and failure of a strategic alliance, partnering, coalition and joint ventures and have consistently identified culture high among the factors (Bamford et al. 2003; Beugelsdijk et al. 2006; Chakravarthy and Lorange 1991). There acceptance of culture as a critical factor for strategic alliances as procurement systems is well acknowledged (see Beugelsdijk et al. 2006; Chakravarthy and Lorange 1991; Harrigan 1988; Lorange and Roos 1992). Murray and Kotabe (2005) identified shared cultural values as key to success, and this promotes information sharing that builds on trust, increases commitment and transparency.

Successful alliances and partnerships are impossible to achieve without cultural similarities and/or strategic complementarities that promotes information and risk sharing (Beugelsdijk et al. 2006; Bamford et al. 2003). The survival alliance and partnering are always in doubt when alliance partners show dissimilar cultures (Beugelsdijk et al. 2006; Rao and Swaminathan 1995). Hence, the survival of alliances and partnering demands cultures should be aligned and be compatible (Rao and Swaminathan 1995). Beugelsdijk et al. (2006) provided a simplification of this and argued that the right balance and fit between organizational cultures of the supply chain partners is essential for alliance and partnering success. This creates the needed environmental context that stimulates sharing of information and communication.

There enough evidence of strategic alliances, partnering, coalition and joint ventures transcending national and international boundaries (Patel 2007; Beugelsdijk et al. 2006; Bamford et al. 2003; Chakravarthy and Lorange 1991). Hence, there is an interplay of both micro- (inter-organizational) and macro- (inter-country) cultural systems, norms, values affecting issues of trust, mutual respect, risk sharing and communication (Beugelsdijk et al. 2006; Sambasivan and Yen 2010). Tuuli et al. (2010) highlighted that macro- and micro-level cultures in alliances and partnering affect the dynamics of control mechanisms in team functions and role performance. Generally, cross-cultural differences (macro-level) have impacted on alliance formation, how they share information and relate (Rao and Swaminathan 1995). The precepts of organizational culture offers further proximal indications underlining organization members' behaviour compared to national culture simply due to the fact that it affords members with an organizational identity and facilitates collective commitments (Sirmon and Lane 2004; Beugelsdijk et al. 2006; Sambasivan and Yen 2010).

Xue et al. (2010), Li et al. (2001) and Sørensen (2002) proved that non-traditional procurement systems exhibit strong organizational cultures compared to conventional procurement systems, and this attribute certainly improves a firm's performance by facilitating internal behavioural consistency. Thus, a strong organisational culture provides stable context for effective communication and information sharing compared to situation of weak cultures in organizations due to inconsistencies in behaviours (Li et al. 2001; Sørensen 2002).

From a theoretical perspective, the social exchange theory posits that parties in an exchange relationship will act in a manner that maximizes their positive outcomes and minimizes their negative outcomes (Whitener et al. 1998). Schlict (2004) identified organizational culture as a leading factor that plays a central role in the exchange relationship. Rothaermel and Deeds (2006) found frequent communication in strategic alliances and coalition allow partnering organisations to create codified routines, policies and procedures for managing the alliance relationship. However in situations of cultural dissimilarity, it seems to impede trust development and create more problems than benefits (Anderson and Weitz 1989; Xue et al. 2010). Sometimes too, there can be the incidence of culture clash which affects communication in strategic alliance and partnering (Elmuty and Kathawala 2001; Bamford et al. 2003). However, continuous interpersonal interactions are deemed to have the propensity to reduce cultural difference, improve the understanding of each other and build a necessary foundation for creating current and future value (Bamford et al. 2003; Xue et al. 2010). Hence, pursuit of understanding culture in alliances, partnering and coalition can help in the building of a strategic alliance relationship and promote the formation of strong network ties (Beugelsdijk et al. 2006; Sambasivan and Yen 2010).

9.4 Social Networks and Its Implication on Communication Outcome Among the Team

Non-traditional procurement systems have been one of the new strategies and innovations in procurement and project management accepted as a response to changing market conditions and competitions in the global construction industry and project delivery landscape (Davis and Walker 2009; Kelly 2011; Mignot 2012). It is generally accepted that these new innovative strategies require an analytical assessment that deals with actor interdependence providing a suitable level of detail and information in relation to the nonlinear, complex, iterative, interactive process, network and relational information that construction projects usually comprise (Pryke 2004; Poleacovschi et al. 2017; Borgatti and Cross 2003).

In the notion of Pryke (2004), these new strategies though considered innovative, it still faces and experiences the conceptualization that construction project is an information intensive activity and environment where a number information exchange networks are created by linkages and relationship. This makes continuous evaluation of the interactions and information sharing by focusing on the communication patterns that develop between participants and organizations to a quantitative analysis of relational data (Poleacovschi et al. 2017; Borgatti and Cross 2003). Social network analysis (SNA) affords an insightful analysis on the communication patterns that develop between participants and organizations (Xie et al. 2010; Abbasian-Hosseini et al. 2017). It has severally been suggested that social network analysis has been a well-known methodology useful in determining the conditions of social structures through the assessment of the relations and interrelationships of a set of actors (Xie et al. 2010; Abbasian-Hosseini et al. 2017).

The work of Penuel et al. (2006) gave credence to the used of SNA and entrenched increase in its usage and popularity in sociology and communication sciences for over three decades. Emmitt and Gorse (2007) and Xie et al. (2010) revealed that construction project environment is a social-based interaction environment, and as such a social communication network can provide useful insight into the high level of role and centrality in the interactions through an analysis of relational data. This is because the social analysis is often employed to evaluate and assess the network relations, identify protuberant patterns, trace information flow and measure the strength of network ties in complex organizational communication networks (Pryke 2004; Penuel et al. 2006). In this regard, the role and centrality became the communication network variables which when analysed reveal the nature of pattern of communication for a project in their networks (Xie et al. 2010; Pryke 2004).

From a questionnaire survey conducted on four typologies of non-traditional procurement systems, namely project alliancing (PA), partnering, integrated project delivery (IPD) and public–private partnerships (PPP), a social network analysis of the communication patterns that develop between participants and organisations was assessed. This followed a similar approach by Xie (2002) and Pryke (2004). By using UCINET 6, the communication variables were evaluated to ascertain the degree of communication centrality between members of the project team and provide a visual representation of network actors by establishing coordinates through multidimensional scaling (see Xie 2002). Lines were used as vectors to represent the linkages that existed between the team members with shorter lines showing strong linkages and longer vectors as weak linkages (Pryke 2004; Xie 2002).

The network was observed based on the frequency of information shared in the role and centrality among the project team and summarized in the communication frequency matrix Table 9.1. The analysis was based on the communication frequency matrix, highlighting communication role, centrality and linkage (after Xie 2002; Pryke 2004).

From this, project team participants are considered close to each other when they communicate frequently with each other. Thus, from the network, the visual representations become very useful in identifying the structure of the network by showing the centrality of each participant as well the nature and strength of their communication. The project teams included project managers, architects, construction managers, engineers (structural/electrical and mechanical), quantity surveyors, contractors and subcontractors and suppliers.

From the results of the SNA analysis, the positional role of each participant in the communication network at the point in time in the four typologies of non-traditional procurement systems (project alliance, partnering, IPD and PPP) were observed. In the case of project alliance arrangements, the results indicated that, the project manager and the architects were the central actors in the communication network with regards to information sharing, relational data and interactions and were well connected to the rest of the project team, demonstrating a significant role in the project

Table 9.1 Communication frequency matrix for project alliance, partnering, IPD and PPP

Communication frequency	Project alliance							
	Project managers	Architects	Quantity surveyor	Construction managers	Engineers	Contractors	Subcontractors	Suppliers
Project managers	0	4	4	2	3	1	1	1
Architects	4	0	4	1	2	3	1	1
Quantity surveyors	4	4	0	1	1	3	2	2
Construction managers	2	1	1	0	1	3	3	2
Engineers	3	2	1	1	0	1	1	1
Contractors	1	3	3	3	1	0	4	3
Subcontractors	1	1	2	3	1	4	0	3
Suppliers	1	1	2	2	1	2	3	0
Total								
Communication frequency	Partnering							
	Project managers	Architects	Quantity surveyor	Construction managers	Engineers	Contractors	Subcontractors	Suppliers
Project managers	0	3	4	3	2	2	1	2
Architects	3	0	4	4	4	4	2	1
Quantity surveyors	4	4	0	4	3	4	3	3
Construction managers	3	4	4	0	2	4	4	4
Engineers	2	4	3	2	0	3	1	1

(continued)

Table 9.1 (continued)

Communication frequency	Partnering							
	Project managers	Architects	Quantity surveyor	Construction managers	Engineers	Contractors	Subcontractors	Suppliers
Contractors	2	4	4	4	3	0	4	3
Subcontractors	1	2	3	4	1	4	0	4
Suppliers	2	1	3	4	1	3	4	0
Total								
Communication frequency	Integrated project design (IPD)							
	Project managers	Architects	Quantity surveyor	Construction managers	Engineers	Contractors	Subcontractors	Suppliers
Project managers	0	3	3	2	2	3	1	1
Architects	3	0	4	2	2	3	2	1
Quantity Surveyors	3	4	0	3	2	3	2	2
Construction managers	2	2	3	0	2	4	2	2
Engineers	2	2	2	2	0	2	1	1
Contractors	3	3	3	4	2	0	3	3
Subcontractors	1	2	2	2	1	3	0	3
Suppliers	1	1	2	2	1	3	3	0
Total								

(continued)

Table 9.1 (continued)

Communication frequency	Public–private partnership							
	Project managers	Architects	Quantity surveyor	Construction managers	Engineers	Contractors	Subcontractors	Suppliers
Project managers	0	3	3	1	1	3	3	2
Architects	3	0	3	2	2	2	1	1
Quantity surveyors	3	3	0	1	2	4	2	2
Construction managers	1	2	1	0	1	1	1	1
Engineers	1	2	2	1	0	1	1	1
Contractors	3	2	4	1	1	0	2	2
Subcontractors	3	1	2	1	1	2	0	2
Suppliers	2	1	2	1	1	2	2	0
Total								

Note Several times per day = 4; daily = 3; weekly = 2; biweekly = 1; not at all = 0 delivery. Interestingly, in spite of the significant roles of the quantity surveyor and contractors in project delivery, they were not in the central position during the project delivery process. This result suggests that the role and centrality of project managers and architects in project alliance in information sharing and interactions among the project team is critical. This is slightly at variance to the insights provided by Xie et al. (2010) where the project manager and quantity surveyor were found to be at the centre of a project alliance relational data and information sharing.

In the case of partnering, from the results, it was observed that architects, quantity surveyors and the contractor were the central actors in the communication network with regard to information sharing, relational data and interactions and were well connected to the rest of the project team, demonstrating their significant role in the project delivery. This suggests that, in the case of partnering project delivery arrangements, the architects, quantity surveyors and the contractor are centrally placed, and thus, they can play a significant role as brokers and gatekeepers to both enable and control the flow of information in the networks.

The IPD somewhat revealed similar results to project alliance in interaction patterns and information sharing among the project team. The results show that after an adjustment to the model by eliminating non-connected nodes, the project manager and the architects were the central actors in the communication network with regards to information sharing, relational data and interactions and were well connected to the rest of the project team, demonstrating their significant role in the project delivery. This is slightly different from the results espoused by Xie et al. (2010) in partnering arrangements.

In the case of PPP, the network analysis indicated a slightly changed pattern where the project manager becomes the central actor and the project team members have moved towards the centre with only the quantity surveyor and subcontractor being away from the central role in interactions and information sharing. From the results, it can be stated that generally across various non-procurement models and typologies, an analysis of the communication frequency matrix focusing on communication role, centrality and linkage, different project team members are the central actors in the communication network with regard to information sharing, relational data and interactions and become the fulcrum for team communication. In this regard, it must be noted that their roles and centrality cannot be overlooked if effective communication can be achieved in specific non-procurement arrangements. This offers a significant implication for multi-team communication in construction project delivery through various non-traditional procurement typologies.

9.5 Summary

Here in this chapter, an insight into teamwork and group dynamics as well as the incidence of culture in teamwork in non-traditional project teams has been provided. Through the use of social network analysis, the nature of relational data and pattern of network in the communication is also highlighted. From this, a clear understanding of the frequency of information shared in the role and centrality among the project team is provided. It can be summarized that though non-traditional procurement systems exhibit consistent and similar characteristics, the results show that the pattern of information flow and sharing in the networks from roles and centrality differ. This revelation could have significant implication on information flow strategies adopted among project teams in various typologies of non-traditional procurement systems in project delivery.

References

Abbasian-Hosseini, S. A., Liu, M., & Hsiang, S. M. (2017). Social network conformity and construction work plan reliability. *Automation in Construction.* https://doi.org/10.1016/j.autcon.2017.01.004.

Anderson, E., & Weitz, B. (1989). Determinants of continuity in conventional industrial dyads. *Marketing Science, 8,* 310–323.

Arrow, H., McGrath, J. E., & Berdahl, J. L. (2000). *Small groups as complex systems: Formation, coordination, development, and adaptation.* Thousand Oaks, CA: Sage.

Bamford, J. D., Gomes-Casseres, B., & Robinson, M. S. (2003). *Mastering alliance structure: A comprehensive guide to design, measure, and organization.* San Francisco, CA: Jossey-Bass.

Belbin, R. M. (1981). *Management teams: Why they succeed or fail* (1st ed.). London: Butterworth-Heinemann.

Belbin, R. M. (1993). *Team roles at work.* Oxford: Butterworth-Heinemann.

Belbin, R. M. (2000). *Beyond the team.* London: Butterworth-Heinemann.

Bell, B. S., & Kozlowski, S. W. J. (2011). Collective failure: The emergence, consequences, and management of errors in teams. In D. A. Hoffman & M. Frese (Eds.), *Errors in organizations* (pp. 113–141). New York, NY: Routledge.

Bell, L. (2001). Patterns of interaction in multidisciplinary child protection teams in New Jersey. *Child Abuse and Neglect, 25,* 65–80.

Beugelsdijk, S., Koen, C. I., & Noorderhaven, N. G. (2006). Organizational culture and relationship skills. *Organization Studies, 27*(6), 833–854.

Bond-Barnard, T. J., Steyn, H., & Fabris-Rotelli, I. (2013). The impact of a call centre on communication in a programme and its projects. *International Journal of Project Management, 31,* 1006–1016.

Borgatti, S. P., & Cross, R. (2003). A relational view of information seeking and learning in social networks. *Management Science, 49*(4), 432–445.

Brown, R. (2000). *Group process: Dynamics within and between groups* (2nd ed.). Oxford: Blackwell.

Chakravarthy, B. S., & Lorange, P. (1991). *Managing the strategy process.* Englewood Cliffs, NJ: Prentice-Hall.

Chan, E. H. W., & Tse, R. Y. C. (2003). Cultural considerations in international construction contracts. *Journal of Construction Engineering and Management, 129*(4), 375–381.

Chen, G., Kanfer, R., DeShon, R. P., Mathieu, J. E., & Kozlowski, S. W. J. (2009). The motivating potential of teams: Test and extension of Chen and Kanfer's (2006) cross-level model of motivation in teams. *Organizational Behavior and Human Decision Processes, 110,* 45–55.

Chen, G., Sharma, P. N., Edinger, S. K., Shapiro, D. L., & Farh, J. (2011). Motivating and demotivating forces in teams: Cross-level influences of empowering leadership and relationship conflict. *Journal of Applied Psychology, 96*(3), 541–557.

Cheung, S. O. (1998). Value of alternative dispute resolution in construction. *Construction Law Journal, 14*(2), 101–110.

Chou, J. S., & Yang, J. G. (2012). Project management knowledge and effects on construction project outcomes: An empirical study. *Project Management Journal, 43,* 47–67.

Crawford, E. R., & Lepine, J. A. (2013). A configural theory of team processes: Accounting for the structure of taskwork and teamwork. *Academy of Management Review, 38,* 32–48.

Dainty, A., Moore, D., & Murray, M. (2006). *Communication in construction-theory and practice.* Abingdon, Oxon: Taylor & Francis.

Davis, P. R., & Walker, D. H. T. (2009). Building capability in construction projects: A relationship-based approach. *Engineering, Construction and Architectural Management, 16*(5), 475–489.

Devine, D. J., Clayton, L. D., Philips, J. L., Dunford, B. B., & Melner, S. B. (1999). Teams in organisations, prevalence, characteristics and effectiveness. *Small Group Research, 30,* 678–711.

Diekmann, J. E., & Girard, M. J. (1995). Are contract disputes predictable? *Journal of Construction and Engineering Management, 121*(4), 355–363.

Elmuty, D., & Kathawala, Y. (2001). An overview of strategic alliances. *Management Decision, 39*(3), 205–218.

Emmitt, S., & Gorse, C. A. (2003). *Construction communication.* Oxford: Blackwell Publishing.

Emmitt, S., & Gorse, C. A. (2007). *Communication in construction teams.* London: Taylor & Francis.

Fellow, R. F., & Hancock, R. (1994). Conflict resulting from cultural differentiation: An investigation of the new engineering contract. In *Council of International Construction Research and Documentation Proceedings on Construction Conflict: Management and Resolution* (pp. 259–267), Rotterdam, The Netherlands.

Friedman, R. A., & Podolny, J. (1992). Differentiation of boundary spanning roles: Labor negotiations and implications for role conflict. *Administrative Science Quarterly, 37,* 28–49.

Gorse, C. A. (2002). *Effective interpersonal communication and group interaction during construction management and design team meetings.* Unpublished PhD. University of Leicester.

Gorse, C., McKinney, I., Shepherd, A., Morley, D. D., & Ellis, R. (2006a). Internal support mechanisms in groups and teams. In *Proceedings of the Association of Researchers in Construction Management. 22nd Annual Conference* (pp. 925–934), September 4–6, University of Central England. ARCOM.

Gorse, C., McKinney, I., Shepherd, A., & Whitehead, P. (2006b). Meetings: Factors that affect group interaction and performance. In *Proceedings of the Association of Researchers in Construction Management. 22nd Annual Conference* (pp. 915–924), September 4–6, University of Central England. ARCOM.

Gorse, C. A., & Sanderson, A. M. (2007). Exploring group work dynamics. In D. Boyd (Ed.), *Proceedings of 23rd Annual ARCOM Conference* (pp. 295–296), September 3–5, 2007, Belfast, UK, Association of Researchers in Construction Management.

Gully, S. M. (2000). Work team research: Recent findings and future trends. In M. Beyerlein (Ed.), *Work teams: Past, present, and future* (pp. 25–44). The Netherlands: Kluwer Academic Publishers.

Gully, S. M., Incalcaterra, K. A., Joshi, A., & Beaubien, J. M. (2002). A meta-analysis of team efficacy, potency, and performance: Interdependence and level of analysis as moderators of observed relationships. *Journal of Applied Psychology, 87*(5), 819–832.

Guzzo, R. A., & Dickson, M. W. (1996). Teams in organizations: Recent research on performance and effectiveness. *Annual Review of Psychology, 47,* 307–338.

Handy, C. (1985). *Understanding organisations* (3rd ed.). London: Penguin.

Harrigan, K. R. (1988). Joint venture and competitive strategy. *Strategic Management Journal, 15,* 291–309.

Hofstede, G. (1991). *Culture and organizations: Software of the mind.* New York: McGraw-Hill.

Hofstede, G. (1997). *Cultures and organization: Software of the mind* (Revised). New York: McGraw-Hill.

Katzenbach, J. R., & Smith, D. K. (1993). *The wisdom of teams: Creating the high performance organization.* Boston, MA: Harvard Business School Press.

Kelly, J. (2011). *Cracking the VFM code: How to identify & deliver genuine value for money in collaborative contracting.* Sydney: Big Fig Publishing Limited, under the imprint of Intelligentsia Press.

Kozlowski, S. W. J., & Bell, B. S. (2003). Work groups and teams in organizations. In W. C. Borman, D. R. Ilgen, & R. J. Klimoski (Eds.), *Handbook of psychology: Industrial and organizational psychology* (Vol. 12, pp. 333–375). London: Wiley.

Kozlowski, S. W. J., & Bell, B. S. (2008). Team learning, development, and adaptation. In V. I. Sessa & M. London (Eds.), *Work group learning* (pp. 15–44). Mahwah, NJ: LEA.

Kozlowski, S. W. J., & Bell, B. S. (2013). Work groups and teams in organizations: Review update. In N. Schmitt & S. Highhouse (Eds.), *Handbook of Psychology: Vol. 12. Industrial and organizational psychology* (2nd ed., pp. 412–469). Hoboken, NJ: Wiley.

Kozlowski, S. W. J., Chao, G. T., Grand, J., Keeney, J., Braun, M., & Kuljanin, G. (2011). Macrocognition and teams: The emergence and measurement of team knowledge. In G. T. Chao & S. W. J. Kozlowski (Chairs), *Macrocognition: The Next Frontier for Team Cognition Research. Symposium presented at the 26th Annual Conference of the Society for Industrial and Organizational Psychology*, Chicago, IL.

Kozlowski, S. W. J., & Ilgen, D. R. (2006). Enhancing the effectiveness of work groups and teams (monograph). *Psychological Science in the Public Interest, 7,* 77–124.

Kumaraswamy, M., & Yogeswaran, K. (1998). Significant sources of construction claims. *International Construction Law Review, 15*(1), 145–159.

Kwofie, T. E., Adinyira, E., & Fugar, F. (2015). An empirical assessment of ineffective communication among mass housing project teams. *Journal of Construction Project Management and Innovation, 5*(2), 1176–1195.

Lee, S. (2005). *The ingroup secure base: An attachment-exploration model of intergroup relations* (Unpublished doctoral dissertation). Cornell University, Ithaca, NY, USA.

Lee, S., & Ling, L. (2007). Understanding affectional ties to groups from the perspective of attachment theory. In E. A. Mannix, M. A. Neale, & C. Anderson (Eds.), *Research on managing groups and teams United Kingdom* (Vol. 10, pp. 217–248). London, UK: Elsevier Science Press.

Lee, S., & Sawang, S. (2016). Unpacking the impact of attachment to project teams on boundary-spanning behaviors. *International Journal of Project Management, 34*(3), 444–451.

Li, J., Lam, K., & Qian, G. M. (2001). Does culture affect behaviour and performance of firms? The case of joint ventures in China. *Journal of International Business Studies, 32*(1), 115–131.

Loosemore, M. (1999). International construction management research cultural sensitivity in methodological design. *Construction Management and Economics, 17*(5), 553–561.

Lorange, P., & Roos, J. (1992). *Strategic alliances.* Oxford: The Blackwell Publishers.

Marmarosh, C. L., & Tasca, G. A. (2013). Adult attachment anxiety: Using group therapy to promote change. *Journal of Clinical Psychology, 69,* 1172–1182.

Marrone, J. A. (2010). Team boundary spanning: A multilevel review of past research and proposals for the future. *Journal of Management, 36,* 911–940.

Mignot, A. (2012). *Who moved my cheese? Adapting to the changing nature of collaboration in infrastructure Alliancing Association of Australasia.* Available from https://www.a3c3.org/. Accessed November 24, 2019.

Murray, J. Y., & Kotabe, M. (2005). Performance implications of strategic fit between alliance attributes and alliance forms. *Journal of Business Research, 58*(11), 1525–1533.

Parkhe, A. (1993). Strategic alliance structuring: A game theoretic and transaction cost examination of interfirm cooperation. *Academy of Management Review, 36*(4), 794–829.

Patel, T. (2007). The role of dynamic cultural theories in explaining the viability of international strategic alliances: A focus on Indo-French alliances. *Management Review, 45*(10), 1532–1559.

Penuel, W. R., Sussex, W., Korbak, C., & Hoadley, C. (2006). Investigating the potential of using social network analysis in educational evaluation. *American Journal of Evaluation, 27*(4), 437–451.

Pett, T. L., & Dibrell, C. C. (2001). A process model of global strategic alliance formation. *Business Process Management Journal, 7*(4), 349–361.

Phua, F. T. T., & Rowlinson, S. (2003). Cultural differences as an explanatory variable for adversarial attitudes in the construction industry: The case of Hong Kong. *Construction Management and Economics, 21*(7), 777–785.

Poleacovschi, C., Javernick-Will, A., & Tong, T. (2017). The link between knowledge sharing connections and employee time savings: A social network analysis. *Construction Management and Economics, 35*(8–9), 455–467.

Pryke, S. D. (2004). Analysing construction project coalitions: Exploring the application of social network analysis. *Construction Management and Economics, 22*(8), 787–797.

Puurtinen, M., Heap, S., & Mappes, T. (2015). The joint emergence of group competition and within-group cooperation. *Evolution and Human Behavior, 36,* 211–217.

Rao, B. P., & Swaminathan, V. (1995). Uneasy alliances: Cultural incompatibility or culture shock. In *Proceedings of the Association of Management 13th Annual International Conference*, Vancouver, Canada.

Rom, E., & Mikulincer, M. (2003). Attachment theory and group processes: The association between attachment style and group-related representations, goals, memories, and functioning. *Journal of Personality and Social Psychology, 84,* 1220–1235.

Rothaermel, F. T., & Deeds, D. L. (2006). Alliance type, alliance experience, and alliance management capability in high-technology ventures. *Journal of Business Venturing, 21*(4), 429–460.

Rowlinson, S. (2001). Matrix organisation structure, culture and commitment—A Hong Kong public sector case study of change. *Construction Management and Economics, 19*(7), 669–673.

Sambasivan, M., & Yen, C. N. (2010). Strategic alliances in a manufacturing supply chain: Influence of organizational culture from the manufacturer's perspective. *International Journal of Physical Distribution and Logistics Management, 40*(5), 456–474.

Schlict, E. (2004). Social culture, corporate culture and exploitation. *Journal of Institutional and Theoretical Economics, 160*(2), 232–240.

Sirmon, D. J., & Lane, P. J. (2004). A model of cultural differences and international alliance performance. *Journal of International Business Studies, 35,* 306–319.

Smith, E. R., Murphy, J., & Coats, S. (1999). Attachment to groups: Theory and measurement. *Journal of Personality and Social Psychology, 77,* 94–110.

Sørensen, J. B. (2002). The strength of corporate culture and the reliability of firm performance. *Administrative Science Quarterly, 47*(1), 70–91.

Sutton, R. I. (2006). The truth about brainstorming. *Business Week*, No. 4002, September, pp. 17–23.

Tierney, W. (1988). Organisation culture in higher education. *Journal of Higher Education, 59*(1), 2–21.

Trompenaars, F. (1993). *Riding the waves of culture: Understanding cultural diversity in business.* London: Economics Books.

Trompenaars, F., & Williams, P. (1999). First-class accommodation, people management. International Personnel Development, April 23, 30.

Tso, J. (1999). Do you dig up dinosaur bones? Anthropology, business, and design. *Design Management Journal*, Fall, 69.

Tuuli, M. M., Rowlinson, S., & Koh, T. Y. (2010). Dynamics of control in construction project teams. *Construction Management and Economics, 28*(2), 189–202.

Vorster, M. C. (1993). *Dispute prevention and resolution.* Blacksburg, VA: Virginia Polytechnic Institute and State University.

Whitener, E. M., Brodt, S. E., Korsgaard, M. A., & Werner, J. M. (1998). Managers as initiators of trust: An exchange relationship framework for understanding managerial trustworthy behavior. *Academy of Management Review, 23,* 513–530.

Wood, J., Wallace, J., & Zeffane, R. M. (2001). *Organisational behaviour: A global perspective.* Australia: Wiley.

Xie, X. (2002). *Communications in construction design.* PhD thesis, Loughborough University, Loughborough.

Xie, C., Wu, D., Luo, J., & Hu, X. (2010). A case study of multi-team communications in construction design under supply chain partnering. *Supply Chain Management: An International Journal, 15*(5), 363–370.

Xue, X., Shen, Q., & Ren, Z. (2010). Critical review of collaborative working in construction projects: Business environment and human behaviors. *Journal of Management in Engineering, 26*(4), 196–208.

Yang, L. R., Huang, C. F., & Wu, K. S. (2011). The association among project manager's leadership style, teamwork and project success. *International Journal of Project Management, 29,* 258–267.

Chapter 10
Improving Communication in Non-traditional Communication: The Case of Critical Competencies

Abstract Communication behaviours have been consistently identified among key soft factors to compliment the technical factors towards improving communication performance in construction project communication. Though various soft factors have been extensively and rigorously pursued, not much has been done of communication competencies and skills for construction communication. The theoretical position of communication performance improvement posits that effectiveness of project team communication appears to be highly dependent on two interrelated factors of which the competences of team leaders and participants to facilitate, stimulate and motivate all members to communicate effectively as a team are most critical. Construction project design and delivery is a collaborative social act and information intensive that relies on effective interaction between project stakeholders relying on key communication skills to effectively communicate. Given the varied meaning and lack of consensus on what constitutes communication competence and skills, this chapter, from a theoretical conception, outlines two key communication competencies and skills for information flow and information composition towards effective task and relational (social) interactions and communication in project delivery. The right application of these skills, will cause project teams and work groups in project delivery to make use of adequate levels for task-based and socio-emotional interactions. This will motivate the right communication performance in construction project delivery.

Keywords Communication behaviour · Communication performance · Task-based interactions · Communication competence

10.1 Introduction

Theoretically, the development of communication skills and competencies has been accepted as one of the significant social actions and human management approaches central to improving communication performance in construction project delivery and cross other communication domain such as management, psychology and general communication. This is considered as the behavioural aspects of communication

T. E. Kwofie et al., *Effective Construction Project Delivery*,
https://doi.org/10.1007/978-3-030-49374-5_10

which has received less attention compared to the technical domain of communication in construction. This chapter explores the key behavioural communication competencies that are essentially critical to engendering effective communication in non-traditional procurement systems. The idea behind this is that these behavioural competencies often relate crucial outcomes in project environment. In this aspect, project participants can develop and entrench their key behavioural competencies in decoding and encoding communication to ensure effective communication performance as they impact team member satisfaction and productivity.

10.2 Skill Development and Acquisition Skills Theories in Construction Communication

In project delivery, it is said that project teams and other participants always face the challenges of figuring out how to effectively communicate an enormous amount of potentially relevant information to ensure effective communication (Henderson 2004; El-Sabaa 2001; Krahn and Hartment 2006). Several studies have identified and underscored the importance and impacts of individual communication competencies for effective team and organizational communication (see Keyton et al. 2013; Henderson 2004, 2008; Rasekh et al. 2012). Meredith et al. (1995) categorized six typologies of skills needed for effective project management and found skills in the area of communication as critical among, organizational, team building, leadership, coping and technological skills. Henderson (2004, 2008) provided a theoretically proven link between team satisfaction and productivity and encoding and decoding communication skills of the project team.

Theoretically, the development of communication skills and competencies has been accepted as one of the significant social action and human management approaches central to improving communication performance in construction project delivery (Marshall-Pointing and Aouad 2005; Kwofie et al. 2014; Kwofie 2015; Kwofie et al. 2015a, b, 2017; Reeta and Neerja 2012; Emmitt and Gorse 2007). Even though communication competence and skills are claimed as very vital to communication effectiveness in project teams, emerging increased virtuality, interdependence, collaborative and multidisciplinary nature of project teams underlines an urging need for desired communication competency behaviours that are crucial for organizing projects efficiently (Kwofie et al. 2015a, b; Reeta and Neerja 2012; Emmitt and Gorse 2007). However, what constitutes communication competence and skills has been very difficult to define and encompassingly conceptualized. From a plethora of studies, it is clear to note that communication competence has been studied from a diverse array of perspective and conceptualization marketing, institutional contexts, conflict, interpersonal relations and intercultural relations (Wilson and Sabee 2003; Wiseman 2002; Jablin and Sias 2001; Spitzberg and Cupach 2002).

Early studies conceptualized communication competence as a form acquired skills and knowledge for an interpersonal influence in communication or an ability to know when, where, how, what and in what manner to communicate (Spitzberg and Cupach 2002; Jablin and Sias 2001; Hymes 1972). However, most of earlier conception of communication competence viewed it as "the speaker-hearer's knowledge of his language" (Chomsky 1965, p. 4). Though this conception is widely held by many studies especially in linguistics and mass communication, it is clear to note that this notion omits performance and lacks the behavioural (interactional) elements and attribute of communication (see Hymes 1972; Cooley and Roach 1984; Salleh 2008; Manusov and Spitzberg 2008; Manusov and Harvey 2001; Henderson 2008).

Salleh (2008) based on Hymes (1972) gave further accentuation to competence as the capability of a person which is dependent on both knowledge and ability for use. By specifying ability as part of competence is an admission that competence allows for the role of non-cognitive factors, such as motivation (Hymes 1972). Hymes (1972) further revealed that knowledge and ability together form the skills required to be deemed as competent. Basically, Hymes suggests that knowledge and ability for use (skill) are required to be perceived competent and that cognition, affection and volition as part of achieving that level. It must be noted that all scenarios presented above strongly related to communication in linguistics, psychology, marketing and other. It must be noted that construction communication differs significantly from these forms of communication.

The skills acquisition theory as applied to communication posits that for one to be deemed competent in communication, there should be declarative knowledge and procedural knowledge which are two interrelated representational systems (Anderson 2005; DeKeyser 2007). The aspect of declarative knowledge refers to static information encoded in memory (DeKeyser 2007). This is also referred to as cognitive knowledge (see Anderson 2005; DeKeyser 2007; Salleh 2008; Ullman 2004). Procedural knowledge on the other hand relates to knowing how to do things, as well as the ability to apply rule-based knowledge to cognitive as well as motor operations (Ullman 2001, 2004; Chein and Schneider 2005; Segalowitz 2010).

It must be noted that this application of the skills acquisition theory relates more to language learning, and it is thus significant to construction communication. This is in the sense that to communicate effectively in construction, one has to gain knowledge that relates to the core functions and role herein termed declarative knowledge, whereas the procedural knowledge will relate to the context and environment of the communication. Louhiala-Salminen and Kankaanranta (2011) operationalized the Chomsky's (1965) seminal approach to communication competence of the linguistic competence theory in business communication and found both knowledge of the language used in the particular communicative situation and an ability to use the language as the determinant of competence.

In the context of construction communication, Kwofie et al. (2015a) explored the attribution theory as applied to communication and task functional and psycho-social communication competencies as the two essential communication behavioural competencies needed by project teams towards effective communication performance. The task functional competencies relate to communication of the core functions and professional roles of the interactants, whereas the psycho-social communication competencies help maintain the social-based interactions and context of the communication (Kwofie et al. 2015b; Salleh 2008; O'Hair et al. 1997; Spitzberg and Cupach 2002). Kwofie et al. (2015b) further alluded that these two behavioural communication competencies are very crucial to engender trust, cohesion and collaboration among project teams through communication.

Inferring from Canale and Swain's (1980) theory, communication competence consisted of grammatical, sociolinguistic, discourse and strategic competence as the four key competencies to effective communication. Grammatical competence related to knowledge of the language for communication, discourse competence represented the knowledge on how to achieve cohesion in communication situations, and strategic competence entailed both verbal and non-verbal communication strategies (Canale and Swain 1980). This has been theoretically supported by Dumitriu et al. (2014) and Peterwagner (2005). From all this, (Hymes 1972) made a valid theoretical assertion that to gain an understanding of an individual's communication competence in any communication context, the individual and his or her interlocutor's competence and the interactional event must be the focal point.

Moving away from the communication competence from the linguistics or language perspective, Cooley and Roach (1984) and McCroskey (1984) conceptualized communication using pedagogical aspect of communication competence and noted cognitive and behavioural dimensions as the two categories noting that communication is situated in the behavioural category. This develops an argument in favour of the theoretical assertion that competence should involve both cognition and behaviour (Cooley and Roach 1984; McCroskey 1984; Wiemann and Backlund 1980). This perspective lends support to Kwofie et al. (2015b) approach to communication competencies using the attribution theory.

Drawing from this argument, Wiemann and Backlund (1980) defined communication competence as the cognitive ability of an interactant in a communication network to suitably choose among available communicative behaviours in order to successfully accomplish the interpersonal goals in the communication process and at the same time maintain the face and line of the other interactant within the communication context.

Cooley and Roach (1984) expanded communication competence and contended that it must always show the demonstration of communication knowledge through the appropriate use of communication skills. From a theoretical perspective, it can be affirmed that, especially in construction communication environment, it can be stated that in construction project environment, the theoretical underpinnings of communication competence lie within the fields of interpersonal and organizational communication within the context of in role functions, teams, organizational context and business and posit that communication behaviours are the main tenets of communication competence (Salleh 2008; O'Hair et al. 1997; Spitzberg and Cupach 2002; Wiseman 2002). Though defining communication competence to embrace the behavioural dimension is generally accepted, it is still noted that, factually, defining and identifying the factors of communication behavioural competencies towards effective communication in any given communication context remain to be difficult (Kwofie et al. 2014, 2015b). From synthesis of the extant literature, one can safely state that there are two main perspectives that seem to advance the development of theory in communication competency scholarships by using cognitive and behavioural perspectives as the predominant themes in competency theories (McCroskey 1984; Spitzberg and Cupach 2002; Spitzberg and Chagnon 2009; Salleh 2008). The cognitive domain conceptualizes communication competency as being a mental phenomenon distinct and separated from behaviour, where the driving and motivating factor towards competency is "knowledge" (Chomsky 1965; Spitzberg 2006; Morreale 2012). This approach is well supported widely and extensively in linguistic, human communication and psychology (Chomsky 1965; Spitzberg and Chagnon 2009).

From the behavioural perspective, the focus is on developing and using appropriate behavioural attributes of the communicators to adapt to the communication tasks as the tenet of communication competence (Hymes 1972; Burleson 2007; Salleh 2008; Morreale 2012). The theoretical foundation of the domain is that communication and communication competency are situated in the behavioural category and that being competent should encompass both cognition (knowledge) and behaviour (attributes) especially in construction communication (Hymes 1972; Wiemann and Backlund 1980; Morreale 2012; Wilson and Sabee 2003; Salleh 2008; Pipsa 2012). This was conceptualized and operationalized by Kwofie et al. (2015b) in Fig. 10.1.

From the work of Henderson (2004, 2008), it can be deduced that the key knowledge and behaviours must relate to encoding and decoding of communicated information in the communication process and network. Given the rapid changes in the global construction industry, complexities of interactions in project teams, work environments and emerging technologies, the need for expanding our knowledge and skills to communication in construction projects that constitute communicative competence is strongly highlighted especially in emerging procurement systems.

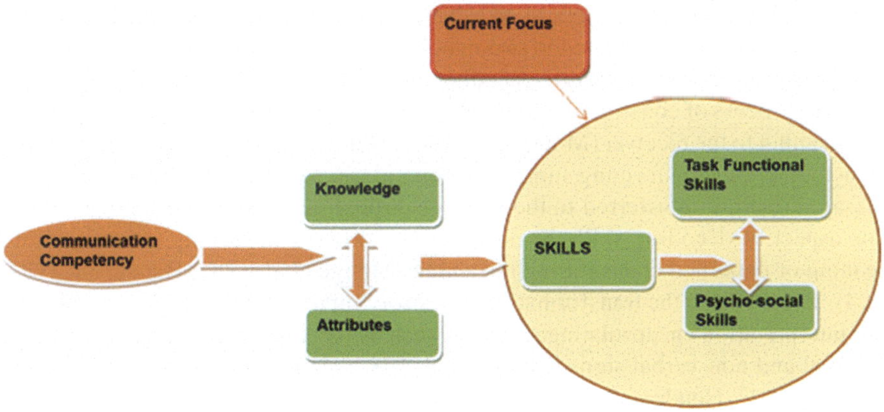

Fig. 10.1 Evolution of components of communication competency in the construction industry (Kwofie et al. 2015b)

10.3 Key Communication Skills for Construction Communication

Developing key communication skills has become a veritable rout to communication competency towards effective communication performance. Marshall-Ponting and Aouad (2005) asserted that theoretically, development of soft skills for communication has become a significant dimension that complements the technical approach to improving communication performance in construction project delivery. Bouchlaghem et al. (2005) also found the development of communication behaviours as useful to ICT-enabled collaborative working environment for concurrent conceptual design. According to Mead (1999) and Otter (2005), the notion gathered was in support of using technology to improve communication performance in construction project delivery. However, their studies gave credence to possessing key soft skills for communication and using these tools as very crucial. As put by Marshall-Ponting and Aouad (2005), that even if tools and technology improve communication, it is humans that operate them; hence, they need to have skills to use them which are vital to the communication process.

From these scenarios, there is no doubt the empirical support to gaining key skills for construction communication. However, the big question which has always been difficult to answer is "what are the key skills vital for communication by project teams?" Attempt to answer this question has been pursued from divergent approaches though theoretically supported. In the fields of project and construction management genre, communication skills are implicit in studies that acknowledge the importance of various organizational aspects of communication to teamwork, team satisfaction, organizational effectiveness and success and various (Henderson 2004; Kwofie et al. 2016, 2017; Ammeter and Dukerich 2002). In Ammeter and Dukerich (2002), listening and persuading behaviours were considered critical skills for project leaders in their interactions. Henderson (2004, 2008) provided an insight into two main skills that related to two major elements of communication especially in construction project delivery—encoding and decoding skills. Encoding skills are related to process of constructing a signal, or message or information that may represent meaning to the receiver (Monge et al. 1982; Henderson 1987). In encoding, there are activities of transforming inner thoughts, ideas, feelings and information into messages that are transferred to the receiver through a medium (Henderson 1987). This aspect of decoding skills involves the the process of turning sensations into meaning or patterned codes through listening (Monge et al. 1982; Henderson 1987, 2004). There is also the transformation of sensory input from the sender into significant interpretation encapsulating activities such as listening, reading and perception of verbal and non-verbal signs (Monge et al. 1982; Henderson 1987; Ammeter and Dukerich 2002; Henderson 2008).

Kwofie et al. (2015b) used the attribution theory and developed a comprehensive communication skill for project teams in construction project delivery which is content driven. From the attribution theory, the position is that the outcome or attribution to communication performance effectiveness is given to both cognitive and behavioural inputs (skills) of the communication process (Zuckerman and Feldman 1984; Salleh 2008; Azemikhah 2005; Bagarić and Djigunović 2007; Morreale 2012). Succar et al. (2013) contended that in any task function performance situation, the competencies of the individual players form the fundamental building blocks of the team and organizational competency. Hence, individuals developing their skills related to tasks in any context will influence higher performance outcome.

From the situation of the attribution theory, it is said that communication performance undoubtedly lies in the "causal locus" domain; thus, the effectiveness of the communication outcome always will depend on the competency input of the communication task performance (communicators) (Peacock 2010; Weiner 2006; Hsieh and Schallert 2008). In other words, a simplified explanation of the theory as applied to communication is that the effectiveness of the communication performance outcome in any communication context is attributed to the communication competency input of the communicators as the causal locus of the success or failure of the communication (Weiner 2006). This was exemplified by Dornyei and Murphey (2003) and Hsieh and Schallert (2008) who found that the causal locus input influencing the communication performance outcome could either be internal or external skills and knowledge of the individual in the communication network in a given context.

From this, Dornyei and Murphey (2003) and Hsieh and Schallert (2008) attributed the internal aspect to the ability possessed by the communicator, whereas the external deals with the communication task environment and the nature of the task. Weiner (2006) provided an in-depth insight which strongly yielded support that the internal causal locus is most critical and has the potential to moderate the impact of external factors if developed effectively. In normal domain performance measures in mainstream human resource genre, several studies have given credence to the shift in focus from knowledge to identification of specific task-related skills and abilities that have the propensity to motivate superior performance outcome (see Azemikhah 2005; Henderson 2004; Mirahmadi et al. 2011). This focus is well supported by the skills acquisition theory which argues in favour of skills ability to offer superior and quicker results because people adapt to skills faster than obtaining knowledge (Mirahmadi et al. 2011).

From the extant literature on communication performance and competencies especially in construction, the arguments have been that for project team participants to be able to engender the needed performance in communication in projects delivery, they must possess two unique behavioural communication skills (Reeta and Neerja 2012; Gorse and Emmitt 2007; Dainty et al. 2006; Salleh 2008; Morreale 2012; Abbasi et al. 2011). These skills are functional task skills and psycho-social skills (Leban and Zulauf 2004; Burleson 2007; Pitts et al. 2012). This is in synch with the theoretical disposition that construction communication lies in the social behavioural domain, and thus given the multidisciplinary nature of construction projects and the complexities with growing virtuality of projects, social communication skills have been useful in stimulating the needed atmosphere for the success of the functional task skills (Leban and Zulauf 2004; Burleson 2007; Pitts et al. 2012). Hence, communication competent project teams must possess two main behavioural communication skills being functional task skills and psycho-social skills.

10.4 Profile of Behavioural Communication Skills for Construction Communication Context

Kwofie et al. (2015b) explored communication competency behaviours among project teams through the use of the attribution theory and affirmed the relevance of functional task skills and psycho-social skills to communication performance outcomes. Kwofie et al. (2015b) following after the framework on the attribution theory developed by Kwofie et al. (2014) deduced that communication behaviours by the project team must be developed along the lines of encoding and decoding in communication information flow and communication information composition. From a synthesis of extensive literature review and mixed approach adopted by Kwofie et al. (2015b), encoding and decoding functional task behavioural skills and psycho-social behavioural skills were deemed very relevant to project team communication effectiveness in construction project delivery. The list of these skills is summarized in Table 10.1

From the dimension of enhancing effective communication outcome, it can be contended that attempt by project teams to adapt and use these communication competency behaviours will stimulate the needed performance in communication across all professional roles and procurement systems. This will be the panacea to overcoming the myriads of communication challenges generally encountered in collaborative, multidisciplinary, cross-functional and interdependent project-based environment for effective communication performance (Kwofie et al. 2015b; Reeta and Neerja 2012; Gorse and Emmitt 2007; Dainty et al. 2006; Salleh 2008).

The development of the functional task communication skills is necessary for encoding the right information related to project tasks as well as enabling receivers of the information to rightly decode and offer appropriate feedback to complete the communication (Kwofie et al. 2014; Gorse and Emmitt 2007). Gorse and Emmitt (2007) revealed that project teams use various forms of communication behaviours for both task and relational interactions during project meetings and discussions. However, the evidence provided by Gorse and Emmitt (2007) shows that project team participants use high levels of task-based interaction and low levels of socio-emotional interaction in construction meetings. In an unchallenged evidence from the use of these behaviours, "the usual adversarial environment often associated with construction was not found, indeed the level of negative emotion and critical discussion was so low that it could be suggested that problems may pass unchallenged" (Gorse and Emmitt 2007, p. 9).

Communication generally in all intent and purposes tend to alter the cognitive and social environment of the persons being addressed (Sperber and Wilson 1986). Hence, the development of the psycho-social skills for both encoding and decoding stands to alter the social context and interactions, thereby ameliorating the tensions in the interactions. This undoubtedly emphasizes the significance of the psycho-social behavioural skills for the project team participants. This is because with the right use of task function skills and psycho-social emotions in encoding the information through a given communication network and medium, it is expected that the thought

Table 10.1 Behavioural communication competency skills for project teams

Functional task behavioural skills	Psycho-social behavioural skills
Functional task communication skills (encoding) • Ability to use language accurately to communicate project-related information very well • Typically gets right to the point precisely in communicating project-related information • Writes efficiently and a way that is easy to understand • Expresses his/her ideas clearly • Not difficult to understand when she/he speaks • Generally says the right thing at the right time • Deliver accurate change request on time • Ability to use specific technical vocabulary and genre conventions correctly • Ability to communicate accurate and complete drawings • Technology-mediated skills *Functional task communication skills (decoding)* • Ability to pay attention to detail of what is being communicated verbally, written and non-verbal • Being a good and effective listener • Ability to ask precise questions to aid understanding of communicated information • Following directions and instructions correctly • Evaluating information accurately to offer accurate feedbacks • Flexibility and tolerance towards others • Technology-mediated skills to access communicated information • Ability to read and understand drawing information	*Psycho-social behavioural skills (encoding)* • Being very sensitive to the needs of others of the moment • Be able to deal with others effectively • Willingness to share required information • Ability to agreeing with team in discussions • Being very polite in communicating information • Ability to share communicated information with honesty • Problem-solving and conflict resolution skills • Show open-mindedness and non-judgmental attitude • Motivation and willingness to communicate with the subordinates • Being assertiveness in communication process • Being trustworthy in communicating with team • Ability to effect goal and ethics-oriented motivation in the communication • Planning and preparing well for meetings, discussions and writings • Adjusting to changing situations • Persuading and engaging team participants • Being cheerful in giving advice and information • Ability to control emotions in discussions and written information • Greeting others in team at meetings and discussions • Ability of showing respect to team members in sharing information • Avoiding offensive language in sharing information • Ability to use non-verbal cues correctly and appropriately *Psycho-social behavioural skills (decoding)* • Is easy to talk to • Usually responds quickly to messages • Cooperating effectively with team members • Showing honesty in information received • Open-mindedness and non-judgmental attitude • Showing trusting ability • Adjusting to changing situations • Avoiding offensive language in giving feedback • Ability to read and understand non-verbal cues • Ability to control emotions in discussions, giving and interpreting information (emotional intelligence) • Greeting others and responding to greetings

Source Adapted after Monge et al. (1982), Keyton et al. (2013), Kwofie et al. (2015a), Henderson (2004, 2008)

process and possible actions of the receiver in the communication loop can be altered positively to ensure effective communication outcome (Gorse and Emmitt 2007; Otter and Emmitt 2007).

Generally, construction project delivery may entail project teams from same organization or different organizations with different cultural and professional backgrounds and goals. It is often possible that this varied background tends to show signs of varied approaches and skills in communication. Hence, the development of these skills can be a sure way to bridging the communication skill gap which can be vital in ensuring effective communication and interactions. This in no doubt will make the ability to communicate accurate and complete project-related information with right social undertones crucial among the project team. Additionally, the right psycho-social behavioural skills can also motivate the needed integration, satisfaction, mutual respect, trust and effectiveness among the team in the delivery process (Spitzberg and Cupach 2002; Salleh 2008; Henderson 2008; Burleson 2007; Keyton et al. 2013). The significance of this is amplified especially in cross-cultural, multidisciplinary interdependent project environments. This tends to reduce various forms of psycho-social barriers in communicating and collaborating across functions and task and across organizations (Reeta and Neerja 2012; Leban and Zulauf 2004).

Dulewicz and Higgs (2000) and Leban and Zulauf (2004) also described psycho-social communication skills as being critical towards integration and team effectiveness in virtual project teams. The empirical account given by Kwofie et al. (2014, 2015b) supports the fact that showing respect, being cheerful, control over emotions, being persuasive greeting and correct use of non-verbal cues are key social skills among the project teams. This revelation gives credence to the significance of certain psycho-social behavioural skills such as respect and trust as being very important in team functions and organizations as suggested by Madlock (2012) and Krahn and Hartment (2006). The import of this is that in many countries especially in developing countries, issues of greetings, respect, trust and verbal cues are elements that are founded of culture and tradition. Given the strong cultural orientation of the construction project environment especially in developing countries, it is enough to draw on good social skills and emotions so as to limit the incidence of arguments, conflicts, mistrust, etc., through the use of right posture, gestures, verbal cues and body language.

10.5 Summary

The need to gain key communication behaviours that can engender the needed competencies for effective communication performance is generally accepted in construction management scholarships, and this is not in doubt. However, what has become a struggle is what skills will constitute the right behaviours for communication competencies among project teams and organizations. Indeed, this chapter has provided a theoretical insight and given credence to the fact that functional task skills and psycho-social skills are two key behavioural skills towards communication competency and effective communication performance outcome. The significance of these skills is embedded in the theoretical foundation that interpersonal and organizational communications lie in the human behavioural domain, and thus, functional task skills and psycho-social skills are two key behavioural skills that account for the "causal locus" of communication performance outcome.

References

Abbasi, M. H., Siddiqi, A., & Ain-Azim, R. (2011). Role of effective communications for enhancing leadership and entrepreneurial skills in university students. *International Journal of Business and Social Science, 2*(10), 242–250.

Ammeter, A. P., & Dukerich, J. M. (2002). Leadership, team building, and team member characteristics in high performance project teams. *Engineering Management Journal, 14*(4), 3–10.

Anderson, J. (2005). *Cognitive psychology and its implications* (6th ed.). New York NY: Worth Publishers.

Azemikhah, H. (2005). *The design of competency based learning resources for VET training packages using learner centred, work centred and attribute focused simulation strategies.* Paper presented at the Australian Vocational Education and Training Research Association Conference, 8th, Brisbane, 2005, Brisbane.

Bagarić, V., & Djigunović, J. M. (2007). Defining communicative competence. *Metodika, 8*(1), 94–103.

Bouchlaghem, D., Shang, H., Anumba, C. J., Cen, M., Miles, J., & Taylor, M. (2005). ICT-enabled collaborative working environment for concurrent conceptual design. *Architectural Engineering and Design Management, 1*(4), 261–280.

Burleson, B. R. (2007). Constructivism: A general theory of communication skill. In B. B. Whaley & W. Samter (Eds.), *Explaining communication: Contemporary theories and exemplars* (pp. 105–128).

Canale, M., & Swain, M. (1980). Theoretical bases of communicative approaches to second language teaching and testing. *Applied Linguistics, 1,* 1–47.

Chein, J., & Schneider, W. (2005). Neuroimaging studies of practice-related change: fMRI and meta-analytic evidence of a domain-general control network for learning. *Cognitive Brain Research, 25*(3), 607–623.

Chomsky, N. (1965). *Aspects of the theory of syntax.* Cambridge, MA: MIT Press.

Cooley, R. E., & Roach, D. A. (1984). A conceptual framework. In R. N. Bostrom (Ed.), *Competence in communication: A multidisciplinary approach* (pp. 11–32). Beverly Hills, CA: Sage.

Dainty, A., Moore, D., & Murray, M. (2006). Communication in construction-theory and practice. Abingdon, Oxon: Taylor & Francis.

DeKeyser, R. (2007). Skill acquisition theory. In B. VanPatten & J. Williams (Eds.), *Theories in second language acquisition: An introduction* (pp. 97–113). Mahwah, NJ: Lawrence Erlbaum Associates.

Dornyei, Z., & Murphey, T. (2003). *Group dynamics in the language classroom.* Cambridge: Cambridge University Press.

Dulewicz, V., & Higgs, M. (2000). Emotional intelligence: A review and evaluation study. *Journal of Managerial Psychology, 15*(4), 341–355.

Dumitriu, C., Timofti, I. C., & Dumitriu, G. (2014). Communicative skills and/or communication competence? *Procedia - Social and Behavioural Sciences, 141,* 489–493.

El-Sabaa, S. (2001). The skills and career path of an effective project manager. *International Journal of Project Management, 19*(1), 1–7.

Emmitt, S., & Gorse, C. A. (2007). *Communication in construction teams*. London: Taylor & Francis.

Gorse, C. A., & Emmitt, S. (2007). Communication behaviour during management and design team meetings: A comparison of group interaction. *Construction Management and Economics, 25,* 1197–1213.

Henderson, L. S. (1987). The contextual nature of interpersonal communication in management theory and research. *Management Communication Quarterly, 1*(1), 7–31.

Henderson, L. S. (2004). Encoding and decoding communication competencies in project management—An exploratory study. *International Journal of Project Management, 22,* 469–476.

Henderson, L. S. (2008). The impact of project managers' communication competencies: Validation and extension of a research model for virtuality, satisfaction, and productivity on project teams. *Project Management Journal, 39,* 48–59.

Hsieh, P. H., & Schallert, D. L. (2008). Implications from self-efficacy and attribution theories for an understanding of undergraduates' motivation in a foreign language course. *Contemporary Educational Psychology, 33,* 513–532.

Hymes, D. (1972). On communicative competence. In J. B. Pride & J. Holmes (Eds.), *Sociolinguistics: Selected readings*. Baltimore: Penguin.

Jablin, F. M., & Sias, P. M. (2001). Communication competence. In F. M. Jablin & L. L. Putnam (Eds.), *The new handbook of organizational communication: Advances in theory, research, and methods* (pp. 819–864). Thousand Oaks, CA: Sage.

Keyton, J., Caputo, J. M., Ford, E. A., Fu, R., Leibowitz, S. A., Liu, T., et al. (2013). Investigating verbal workplace communication behaviors. *Journal of Business Communication, 50*(2), 152–169.

Krahn, J., & Hartment, F. (2006). *Effective project leadership: A combination of project manager skills and competencies in context*. Paper presented at the Biennial Meeting of the Project Management Institute Research Conference, Montreal, Canada.

Kwofie, E. (2015). *Contribution of unique features of mass housing projects to project team communication performance* (Unpublished Ph.D. thesis). Kwame Nkrumah University of Science and Technology, Department of Building Technology, Kumasi.

Kwofie, T. E., Fugar, F., & Adinyira, E. (2015a). Contribution of multiple construction site management features to project team communication effectiveness: The case of mass housing projects. *Engineering Project Organization Journal, 5,* 180–193.

Kwofie, T. E., Fugar, F., & Adinyira, E. (2015b). Identifying the communication competency behaviors of mass housing project teams in developing countries: An exploratory study. *Covenant Journal of Research in the Built Environment (CJRBE), 3*(1), 22–43.

Kwofie, T. E., Fugar, F., & Adinyira, E. (2016). Modelling the effect of housing design unit contract packaging on mass housing project team communication performance. *Journal of Construction in Developing Countries, 21*(1), 35–50.

Kwofie, T. E., Adinyira, E., & Fugar, F. (2017). Theoretical and practical implications for engendering project team communication effectiveness in mass housing project delivery in Ghana. *Journal of Engineering, Design and Technology, 15*(6), 826–844.

Kwofie, T. E., Fugar, F. D. K, Adinyira, E., & Ahadzie, D. K. (2014). A conceptual framework for evaluating communication performance among mass housing project team. In O. Ejohwomu & O. Oshodi (Eds.), *Proceedings of 3rd International Conference on Infrastructure Development in Africa* (pp. 6–21), March 17–19, 2014, Abeokuta, Nigeria.

Leban, W., & Zulauf, C. (2004). Linking emotional intelligence abilities and transformational leadership styles. *Leadership & Organization Development Journal, 25*(7/8), 554–564.

Louhiala-Salminen, L., & Kankaanranta, A. (2011). Professional communication in a global business context: The notion of global communicative competence. *IEEE Transactions on Professional Communication, 54*(3), 244–262.

Madlock, P. E. (2012). The influence of conflict management, leadership, and communication on employee job satisfaction. *Human Communication. A Publication of the Pacific and Asian Communication Association, 15*(2), 121–138.
Manusov, V., & Harvey, J. H. (2001) *Attribution, communication behavior, and close relationships.* Cambridge: Cambridge University Press. 978-0-521-77089-7.
Manusov, V., & Spitzberg, B. H. (2008). Attributes of attribution theory: Finding good cause in the search for theory. In D. Braithwaite & L. A. Baxter (Eds.), *Engaging theories in interpersonal communication* (pp. 37–50). Thousand Oaks, CA: Sage.
Marshall-Ponting, A. J., & Aouad, G. (2005). An nD modelling approach to improve communication process for construction. *Automation in Construction, 14*(3), 311–321.
McCroskey, J. C. (1984). Communication competence: The elusive construct. In R. N. Bostrom (Ed.), *Competence in communication* (pp. 259–268). Beverly Hills, CA: Sage.
Mead, S. T. (1999). *Communication effectiveness in intranet based construction projects* (Unpublished Ph.D. thesis). Department of Civil and Building Engineering, Loughborough, University, Loughborough, UK.
Meredith, R., Posner, B. Z., & Mantel, S. J., Jr. (1995). *Project management: A managerial approach.* New York: Wiley.
Mirahmadi, H., Jalilzadeh, K., & Nosratzadeh, H. (2011). Skill acquisition theory in second language teaching: A focus on productive skills. In *International Conference on Languages, Literature and Linguistics IPEDR* (Vol. 26) © (2011). Singapore: IACSIT Press.
Monge, P. R., Bachman, S. G., Dillard, J. R., & Eisenberg, E. M. (1982). Communication competence in the workplace: Model testing and scale development. In M. Burgoon (Ed.), *Communication yearbook* (Vol. 5, pp. 505–528). USA: Transaction Books.
Morreale, S. P. (2012). Competent and incompetent communication. In *21st century communication: A reference handbook* (pp. 445–454). Thousand Oaks, CA: SAGE.
O'Hair, D., Friedrich, G., Wiemann, J., & Wiemann, M. (1997). *Competent communication* (2nd ed.). USA: St. Martin's Press.
Otter, A. F. H. J. (2005). *Design team communication using a project website* (Ph.D. thesis). Bouwstenen 98, Technische Universiteit Eindhoven, Eindhoven.
Otter, A., & Emmitt, S. (2007). Exploring effectiveness of team communication balancing synchronous and asynchronous communication in design teams. *Engineering, Construction and Architectural Management, 14*(5), 408–419.
Peacock, M. (2010). Attribution and learning English as a foreign language. *ELT Journal, 64*(2), 184–193.
Peterwagner, R. (2005). *What is the matter with communicative competence?* Wien: LIT.
Pipsa, P. (2012). *Interpersonal communication competence and collaborative interaction in SME Internationalization.* Jyväskylä. Jyväskylä University Printing House. ISBN 978-951-39-4701-9.
Pitts, V. E., Wright, N. A., & Harkabus, L. C. (2012). Communication in virtual teams: The role of emotional intelligence. *Journal of Organizational Psychology, 12*(3/4), 21–34.
Rasekh, A. E., Zabihi, R., & Rezazadeh, M. (2012). An application of Weiner's attribution theory to the self-perceived communication competence of Iranian intermediate EFL learners. *Elixir International Journal of Psychology. Elixir Psychology, 47,* 8693–8697.
Reeta, R., & Neerja, P. (2012). Communication competence of Indian engineers in IT & ITeS sector. *Indian Journal of Industrial Relations, 47*(3), 62–78. Shri Ram Centre for Industrial Relations and Human Resources. ISSN: 0019-5286.
Salleh, L. M. (2008). Communication competence: A Malaysian perspective. *Pacific and Asian Communication Association, 11*(3), 303–312.
Segalowitz, N. (2010). *Cognitive bases of second language fluency.* London: Routledge.
Sperber, D., & Wilson, D. (1986). *Relevance communication and cognition.* Oxford: Blackwell.
Spitzberg, G. H. (2006). Preliminary development of a model and a measure of computer-mediated communication (CMC) competence. *Journal of Computer-Mediated Communication, 11,* 629–666.

Spitzberg, B. H., & Chagnon, G. (2009). Conceptualizing intercultural communication competence. In D. K. Deardorff (Ed.), *The SAGE handbook of intercultural competence* (pp. 2–52). Thousand Oaks, CA: Sage.

Spitzberg, B. H., & Cupach, W. R. (2002). Interpersonal skills. In M. L. Knapp & J. A. Daly (Eds.), *Handbook of interpersonal communication* (pp. 564–612). Thousand Oaks, CA: Sage.

Succar, B., Sher, W., & Williams, A. (2013). An integrated approach to BIM competency assessment, acquisition and application. *Automation in Construction, 35,* 174–189.

Ullman, M. (2001). A neurocognitive perspective on language: The declarative/procedural model. *Nature Reviews Neuroscience, 2*(10), 717–726.

Ullman, M. (2004). Contributions of memory circuits to language: The declarative/procedural model. *Cognition, 92*(1–2), 231–270.

Weiner, B. (2006). *Social motivation, justice, and the moral emotions: An attributional approach.* Mahwah, NJ: Lawrence Erlbaum Associates.

Wiemann, J. M., & Backlund, P. (1980). Current theory and research in communicative competence. *Review of Educational Research, 50,* 185–198.

Wilson, S. R., & Sabee, C. M. (2003). Explicating communicative competence as a theoretical term. In J. O. Greene & B. R. Burleson (Eds.), *Handbook of communication and social interaction skills* (pp. 3–50). Mahwah, NJ: Lawrence Erlbaum.

Wiseman, R. L. (2002). Intercultural communication competence. In W. B. Gudykunst & B. Mody (Eds.), *Handbook of international and intercultural communication* (2nd ed., pp. 207–224). Thousand Oaks, CA: Sage.

Zuckerman, M., & Feldman, L. S. (1984). Actions and occurrences in attribution theory. *Journal of Personality and Social Psychology, 46*(3), 541–550.

Chapter 11
Exploring Information and Communications Technology for Enhanced Communication in Non-traditional Procurement

Abstract In today's digitalization era, the global construction industry has witnessed a naissance Information Communication Technology (ICT) tools and platforms for which have had significant impact on the way project teams and organizations interact as well as having impact on storing, processing, retrieving, disseminating and sharing of information. The information-intensive nature of construction project delivery process requires real-time access and flow of information between all the project team members. Given the present scenario, when project participants are geographically separated and with increased virtuality, adoption of ICT that enables effective communication is critical. However, despite the wave of enthusiasm for many ICT tools such as intranet, Internet, extranet, project Website, BIM, Internet of things, augmented and virtual reality as a platform for information sharing and interactions, among project teams, issues of their performance and information-sharing behaviours across various procurement typologies are yet to be explored as a strategic tool towards effective communication. In the current state of the industry, there is also growing recognition of the need to understand the ICT needs of the individuals and how they use these tools to communicate within project teams to engender effective communication. Additionally, project team participants come from different organizations and perform varied roles using varieties of information systems and tools, having different rates of adoption of the available communication tools as well as preferences for specific means of communication and information sharing. In the current digitalization era, the effectiveness of the project team information sharing and interactions is dependent on clear understanding of the performance of specific ICT communication media and tools adopted across various procurement typologies and their information-sharing behaviours. This chapter explores the evolution of ICT tools and platforms for information sharing and interaction in construction project delivery in non-traditional procurement models. This knowledge will enable the strategic adoption of ICT tools for information sharing and interactions among project teams in construction project delivery. By this strategic drive, the use of these ICT tools can aid in improving collaborative working and integration of project participants which is proving to be elusive in practice especially in conventional procurement models.

T. E. Kwofie et al., *Effective Construction Project Delivery*,
https://doi.org/10.1007/978-3-030-49374-5_11

Keywords Information communication technology · Effective communication · ICT tools · Information sharing

11.1 Introduction

Given the well accepted impetus ICT tools give to construction communication effectiveness and performance, the industry continues to explore various tools that can stimulate effective communication in construction project delivery across various procurement systems. The combination of both the technical dimension and soft human behavioural skills as the route to effective communication performance in construction is well supported. In this chapter, a clear focus into technical dimension of communication by exploring the typologies of ICT tools is pursued. ICT-aided communication can facilitate interactions and information sharing among colocated, geographically distributed and virtual project team participants across the life cycle phases of project delivery.

11.2 Exploring Key ICT Tools for Enhanced Communication in Non-traditional Procurement Systems

Technical tools including ICT tools are among the technological factors that can induce improved communication performance in construction project delivery (Marshall-Ponting and Aouad 2005; Yang et al. 2007; Ahuja et al. 2010; Thorpe and Mead 2001). Even though generally the consensus is that a combination of both technology and soft human management approaches brings about the needed changes to construction communication effectiveness, the role of ICT from the technological dimension is very significant. The construction industry is said to be slow to ICT adoption compared to other industry (Yang et al. 2007; Ahuja et al. 2010). Such benefits as effective decision-making, faster and easier access to project information, improved communication, integrated relationship, improved information flow and improved management of virtuality of teams are inherent from ICT adoption to construction project delivery (Love et al. 2004; Root and Thorpe 2001; Yang et al. 2007; Bouchlaghem et al. 2005a).

Early application of ICT in construction was primarily limited to design and construction activities such as drawings and planning (scheduling) activities (Yang et al. 2007; Love et al. 2004; Thorpe and Mead 2001). Technical, managerial, social, cultural, political and economic reasons have been ascribed to the slow pace of ICT diffusion and uptake in construction especially in the area of communication (Yang et al. 2007; Ahuja et al. 2010; Peansupap and Walker 2005; Xie et al. 2018). In spite of the significant strides and development in ICT tools and technology for construction information transfer and communication, much of the interactions, building and information transfer is still dominated by the conventional human communication in printed drawings, letters and other documents especially in developing countries (Ahuja et al. 2009; Yang et al. 2007; Gorse and Emmitt 2007; Emmitt and Gorse 2007).

Insight espoused by Ahuja et al. (2010), Thorpe and Mead (2001) lends theoretical, practical and contextual support and linkage to the fact that IT-enhanced communication protocols for building project management have had significant impact on communication, information sharing and interactions among project teams in construction project delivery. Suffice to say that ICT support for construction communication, interactions, information sharing and management has been an active research area for the past three decades. Notable works include groupwares, intranet, Internet, emails, video conferences, messaging and texting, project Websites and 3D computer-aided systems (see Shen 1992; Mead 1999; Rojas and Songer 1999; Otter 2005; Lou et al. 2005; Bouchlaghem et al. 2005a; Duan and Zhou 2006; Ahuja et al. 2010; Rad et al. 2010; Nitithamyong and Skibniewski 2004; Dawood et al. 2002). In recent times, the focus has been on developing information-sharing platforms in 3D and 4D as well as dealing with virtuality and digitalization in information sharing leading to tools such as BIM (Rad et al. 2010; Leidner 2010; Egbu and Robinson 2005; Nitithamyong and Skibniewski 2011; Akponeware and Adamu 2017; Liu et al. 2017; Hosseini et al. 2012; Xie et al. 2018).

Generally, the communication environment in construction has changed considerably through the application of ICT tools by exploring tools for automatic information sharing, reengineering communication networks and widening available communication media (Liu 2009; Xie et al. 2018). However, the general IT productivity paradox shows that investments in IT do not always result in higher productivity and performance (Brynjolfsson 1993; Brynjolfsson and Hitt 1998; Martinsons and Martinsons 2002). From the turn of the 1990s, the development has focused on ICT as a medium for construction communication that is capable of establishing favourable supply chain relationships along four main lines of taxonomies including standardization, visualization, communication and integration (Thorpe et al. 1998; Mead 1999). The ICT tools are summarized along the line of the taxonomies in Table 11.1.

According to Mead (1999), this taxonomy offers a useful way of classifying current information technologies, and it must also be acknowledged that some of the technologies overlap in their usage across other classifications. For example, though email is classified as a communication tool, it is also often used as an integrated medium for the transfer of electronic files (Mead 1999). Same can be said of EDI which is purely a standardization technology, but it also encompasses elements for communication and integration in its use (Thorpe et al. 1998; Mead 1999).

Table 11.1 Taxonomy of construction information technologies

Classification (group)	Types of technology
Standardization technologies	Electronic data interchange (EDI) Bar coding TCP IIP
Visualization	Virtual reality Computer simulation Computer-aided design (CAD) Multimedia
Communication	Electronic mail Video conferencing Project intranets Groupware
Integration	Infobases Local and wide area networks Object-oriented databases Expert systems

Source Thorpe et al. (1998), Mead (1999), Thorpe and Mead (2001), Graham et al. (1996), Otter (2005), Liu (2009), Xie et al. (2018)

11.2.1 Groupwares for Project Collaborative Information Sharing

Groupwares are classified as a computer-based system enabling technology that support groups of participants who are engaged in a common task by providing an integrated interface to a shared environment (Wilson 1991; Ellis and Wainer 1994; Mead 1999). According to Mead (1999), groupwares as an enabling technology have been essentially developed along the lines of asynchronous and synchronous or real-time groupware. The asynchronous systems allow project teams to use the tools at different times at different places, and this includes email, newsgroups, workflow systems, group calendars, and collaborative writing systems (Mead 1999; Dawood et al. 2002; Otter 2005; Otter and Emmitt 2007). The synchronous groupware systems allows team participants in the communication chain and network to use the tools at the same time and could be at different places and this includes shared whiteboards, video communications, chat systems, and decision support systems (Mead 1999; Otter 2005; Otter and Emmitt 2007).

In the considered experience of Baldwin et al. (1996), the adoption of groupwares and other information technologies in construction communication and information sharing provided a symbiotic benefit to the construction process by facilitating high-speed information exchange and enhancing project teamwork.

11.2.2 Intranet and Internets in Project Information Management and Communication

Mead (1999) defined intranet as a restricted Internet network that utilizes series of linked web pages to manage and distribute related information to individuals of a project team within an organization or inter-organization. In some scenario, the intranet system can be oriented towards participants outside the organization or project teams, and this is referred to as an extranet (Otter 2005; Mead 1999; Otter and Emmitt 2007). In its predominant usage, the term intranets have often been associated with restricted Internet networks using the World Wide Web (WWW) to give team participants and project organizations easy and immediate access to virtually any sort of electronic document (Otter 2005; Xie 2002). With this, project teams and organizations have access to the same electronic information to the project team members regardless of platform or geographic location (Mead 1999; Otter and Emmitt 2007). It must be noted that though project intranets have dominantly been used primarily as a communication enhancement tool, Otter (2005) and Mead (1999) contend that they have also be put to use as an integrating technology.

In construction project delivery, commonly, an Internet-based Web server which stores project information such as specifications, cost elements, design documents, and information requests has been used for project intranets. This allows for project team participants to have access to the information through the use of the Internet-based World Wide Web (WWW) system from anywhere they are located (Otter 2005; Mead 1999; Xie et al. 2010). Any form of changes or modifications to the information is updated and stored on the server, and participants have access to the new information from anywhere they are (Mead 1997, 1999). The adoption or use of project intranet can significantly lead to improved teamwork, streamlined project schedules and enhance accuracy and timeliness in project information access. Project intranet provides a means of effectively managing submittals, requests for information, schedules, specifications, meeting minutes and digital photographs (Mead 1999).

The use of the Internet as an information sharing and communication tool took a boom from the year 2000 and provided construction groups with new avenues for business process engineering (Bouchlaghem et al. 2005a; O'Brien and Al-Soufi 1994; Basu 1996; Mead 1999). In 1968, the Internet evolved from being a group of networks developed by the United States Department of Defense to aid spread the load of several large research computers to other computers around the country (O'Brien and Al-Soufi 1994; Mead 1997, 1999). It involved a series of interconnected computer networks that provided a secure means of communication during a nuclear attack (Mead 1999). The internetworking protocol (IP) is used as a simple computer language for the Internet to move electronic files to remote locations via a series of distributed networks (Vanier and Turk 1995). The operation of the Internet is on protocols that means client machines which are able to send, receive and make request for electronic information from server machines (Vanier and Turk 1995; Mead 1999).

In this case, the servers became what was termed as information brokers which can be accessed more than one client at a time (Vanier and Turk 1995; Mead 1999). By following, the established electronic protocols allow users (clients) and servers to exchange electronic information through the network that connects computers around the world (Vanier and Turk 1995). The use of the Internet as electronic commerce or for communication and information-sharing tool has increased remarkably by which the construction industry has tapped its potential (Thorpe et al. 1998; Mead 1999; Vanier and Turk 1995). The growth of the Internet was facilitated by the rapid development of the World Wide Web (WWW), which created the vehicle for the seamless exchange of text, graphic, audio and multimedia files (Vetter et al. 1994). In construction project delivery or environment, the Internet has acted as a business process engineering of construction business that enables the sharing of electronic data helping to bridge the virtual gap given that clients can access information wherever they are.

Today, the Internet has helped to shape construction business process and information sharing where Internet-based information technology tools have been created as "global offices", allowing local firms to access project information all day round (Baldwin et al. 1996; Mead 1999; Mitchell 1993; Duke and Anumba 1997). Today, it can be said that the Internet has become an invaluable ICT tool for the construction industry without which it is near impossible for construction business to function efficiently. T\It has indeed become a catalyst for improving project team communication and also helped create a fully integrated data processing as well as telepresence to create electronic collaborative environments (Duke and Anumba 1997; Vanier and Turk 1995). Mead (1999) alluded that the use of the Internet has offered project team participants to share project-related word-processed documents, specifications, change orders, CADD drawings, database records and hypertext documents timely and efficiently.

11.2.3 Project Websites (PWS) for Project Delivery

Project Websites are defined as a protected Internet environment often referred to as extranet that is accessible for registered users, with information vaults controlled by a central database (Otter 2005; Love et al. 2001; Ahuja et al. 2009). Project Websites were discovered from the drive to explore the use of new and emerging technologies for design and project team communication (Otter 2005; Wang and Love 2012; Wong and Zhang 2013; Skibniewski and Zavadskas 2013). According to Otter (2005), project Websites were equipped with promising features that were capable of solving important barriers in team communication. Project Websites are generally described as asynchronous means of communication tool which allows for project team participants or clients to access its content at different times and from different places (Otter 2005; Wang and Love 2012; Wong and Zhang 2013). With the aid of Internet functionality, project team participants are allowed an easy access to the project-related information and other documents on PWS through password and identity codes that are linked to specific user rights (Otter 2005; Wong and Zhang 2013; Skibniewski and Zavadskas 2013).

PWS use must be able to control user rights and maintenance through a PWS administrator. The operation of a PWS allows participants according to the appointed database structure with the right status and version to document one's own information stored and updated in PWS as the sender of the information and then facilitates the possibility of viewing or reading, updating and reusing by other team members being the receivers at the times needed (Otter 2005; Wong and Zhang 2013; Skibniewski and Zavadskas 2013). Receivers are allowed a possibility of offering feedback to the information sent by the sender on a PWS by updating the stored information of the sender and by sending a PWS message to the creator of the stored information (Otter 2005; Otter and Emmitt 2007).

Project Websites system allows for information storing, maintaining, updating and viewing information as well as given feedback to project team members. By this, it is said that communication has taken place among the project team (Otter 2005; Otter and Emmitt 2007). Otter (2005) noted that PWS packages generally offer different features which range from basic features for storage, viewing or reading and updating through features for status and version control, with virtual office functions. It also has enhanced superior features for management information, automated procedures for information handling of the users in the form of recording, assembling, sorting and classifying the meta-data of the PWS which is far superior to that of MS Outlook (Otter 2005; Sutton 1996; El-Ghandour and Al-Hussein 2004).

Normally, a PWS system can be implemented centrally at each project team level or distributed in each collaborating project organization level whereby the central tool approach will require an IT-facilitating organization to manage and supervise the use of the PWS (Otter 2005). The work of Otter (2005) revealed that PWS have *Interactive* mode, *Effective* mode, *Active* mode and *Reuse* mode as the communication modes which show that time frequency of information sharing using PWS. This attribute is important for effectiveness of team communication and improvement of team performance (Otter and Emmitt 2007). In the *Interactive* mode, there is the need for a high frequency of updating of actual information allowing for information to be stored and uploaded every hour to register all changes in information and maximize tuning between design team members (Otter 2005). This also demands frequent feedback by team members by updating the information or by sending messages. The use of PWS in the *Interactive* mode, according to Otter (2005), is highly effective for team communication and can substantially improve team performance.

The use of PWS in the *Effective* mode allows for storing and updating actual information within a time frequency of every 4 h to a maximum of one day (Otter 2005; Otter and Emmitt 2007). The use of PWS in the *Effective* mode enhances effective team communication and improves team performance due to the up-to-date status and overview of stored information (Otter 2005; Otter and Emmitt 2007). The *Active* mode takes care of storage of provisional or semi-final information with a time frequency longer than one day and mostly without updating. In this mode, it is accounted that feedback is not expected because most likely, the actual process of storing and updating of information took place in the shared project disc (Otter

2005). The *Reuse* mode deals with the storage of the finalized information of a design process or parts of it to be used in other phases of the design project or future design projects which is often referred to as an electronic or digital library (Otter 2005; Otter and Emmitt 2007).

11.2.4 3D Computer-Aided Systems

The 3D information and communication technology (ICT) is deemed as a response to today's era of digitalization and managing the amount of information involved in construction (Egbu and Robinson 2005; Hosseini et al. 2012). However, the literature is replete with evidence of IT productivity paradox, which shows that investments in IT do not always result in higher productivity (Brynjolfsson 1993; Brynjolfsson and Hitt (1998); Nitithamyong and Skibniewski 2004). Meanwhile, Brynjolfsson (1993), Brynjolfsson and Hitt (1998) and Dehning et al. (2004) iterated that the effective use of IT tools in construction communication varies according to the management of information and technology, lags due to learning and adjustment effects, management and resistance to change, organizational slack, diffusion delay and changes in workflow to use IT effectively.

In recent times, 3D and 4D ICT communication tools have gained credence and have had a deep impact on how construction activities are done through access to accurate information (Rad et al. 2010; Ahuja et al. 2010; Love et al. 2004). In terms of information sharing, the construction industry has witnessed the development of various user-friendly ICT tools and software facilities that are utilized across different stages of construction that receive and process information (data) into different formats for different uses (Peansupap and Walker 2005; Alshawi et al. 2009; Ahuja et al. 2009, 2010). Rad et al. (2010) intimated that these emerging 3D and 4D computer-aided ICT tools make visualization possible and thus improve the aspect of communication and information sharing in construction project delivery.

The construction industry has explored various 3D ICT tools which have primarily applied to enhance distributed organizational interactions and achieve good coordination, collaboration and communication between distributed and virtual project teams (Perry and Sanderson 1998; Wikforss and Lofgren 2007). A developed multi-agent system to support collaborative design in 3D virtual worlds developed by Lou et al. (2005) provided evidence of effectively facilitating the level of communication which was not readily available in conventional computer-aided design (CAD) systems. The effort by Duan and Zhou (2006) saw the creation of computer-supported collaborative design (CSCD) system that enhanced the coherence of product information in the design process through integrating CAD system with product data management (PDM) system. The application of peer-to-peer network for real-time online collaborative computer-aided design by Chen and Tien (2007) proved that accessibility and flexibility in collaborative design are improved.

From the height of these technological innovations in ICT applied to construction communication and information sharing, El-Saboni et al. (2009) still affirmed that computing and communication technologies continue to be of significant importance and impact on improving design communication. In Bråthen and Moun (2016), a 3D BIM kiosk which was introduced to site workers made it possible to get on-site access to up-to-date design information of the planned building. This system allowed for the 3D visualization of complex spatial situations, thereby enhancing understanding, interactions and decisions on the project (Bråthen and Moun 2016). Today, Building Information Modelling (BIM) has become a dominant 3D ICT tool used in the construction industry. The most widespread use of BIM today is in the design and construction phases of construction projects (Jensen and Jóhannesson 2013; Eastman et al. 2008, 2011; Arayici et al. 2012; Miettinen and Paavola 2014). BIM has become an inseparable ICT tool that facilitates digital, machine-readable documentation of building, its performance, its planning, its construction and later its operation (Eastman et al. 2008).

According to Arayici et al. (2012) and Eastman et al. (2011), the required integration and collaboration required in construction process and the industry can be certain only when the processes are underpinned by verified effective information and knowledge sharing which is facilitated by the use of information and communication technology (ICT). In the opinion of Eastman et al. (2011) and Arayici et al. (2012), BIM as a tool can be regarded as an innovation underpinned by digital and virtual 3D communications tool and model that has revolutionized the extent of integration and collaboration within the project delivery process. Cao et al. (2017) found empirical support for the ability of BIM to enhance the ability to enhance collaboration and integration in tasks and information sharing.

Other ICT tools and platforms such as augmented reality and virtuality reality have also become innovative ICT models that have been implemented in the construction industry (see Sampaio 2018; Chen et al. 2007, 2011, 2015; Park et al. 2015; Stanney and Hale 2014; Wang et al. 2014). Virtual reality (VR) and augmented reality (AR) create a computer-generated simulation of 3D environment which allows users (project team participants) to view, access and manipulate the contents of that environment (Chen et al. 2011, 2015). By the access of this environment, a clearer understanding of the project is given as well as access to accurate information on the related project under control (Chen et al. 2007; Park et al. 2015; Stanney and Hale 2014). Chen et al. (2015) recounted that the application of VR and AR has one benefit of design improvement and enhanced communication among project teams by allowing for the user to see a project in its completed form overlaid onto an empty field that improves understanding through visualization.

Beyond the realm of 3D modelling, 4D modelling has become a more recent and enhanced a tool to improve construction collaboration, interactions and access to accurate information (Pittard and Sell 2016; Hardin and Mccool 2015; Lin et al. 2011; Zhang et al. 2018). 4D modelling as an emerging package of BIM *level 2* has become effective to improving health and safety on construction site through the access to accurate information as well as enhancing the understanding of these information. Pittard and Sell (2016) proved that BIM level 2 4D provides platform that creates consistent and structured information and management which remains a key factor for reducing both project risk and wasteful processes.

There is no doubt the benefits of ICT tools to construction interactions, information sharing and management. However, it must be noted that the mere adoption of these tools should not be seen as a panacea to improvement. Hence, it is important for project teams to first explore the context suitability and professional role challenges to these tools as it can be a veritable approach to legitimizing and optimizing their benefits to construction project delivery and management.

11.3 The Reality of Virtuality on Collaborative Media *Richness* in Construction Communication

Construction project teams across various procurement typologies and systems are predominantly becoming hybrid in nature in which interactions and communications are increasingly more virtual and less of face-to-face (Ahmed et al. 2017; Hosseini et al. 2017; Trautsch 2003; Manninen 2003). Virtuality has become a key determinant within interactions and information-sharing tasks of project teams within construction organizations (Kim 2009; Singh et al. 2011; Chen et al. 2014; Ahmed et al. 2017). The benefits and case for virtual project teams are well acknowledged in the literature (see Trautsch 2003; Bouchlaghem et al. 2005b; Chen et al. 2011; Golparvar-Fard et al. 2009). Golparvar-Fard et al. (2009) enumerated cultural differences, lack of non-verbal cues, lack of trust, difficulties in building relationships and uncertainty in team members' competence, challenges related to communication technology and lack of language competence as the most cluster of challenges to the incidene of virtuality in construction project delivery.

Growing virtuality of construction project teams means that, project leaders and managers across various procurement typologies and systems are leading global teams that work synchronously or asynchronously from various locations thus making decoding and encoding media richness crucial to their interactions (Ahmed et al. 2017; Chen et al. 2011). Communication and interactions among project teams in the wake of virtuality have become very complex and often a challenging process as their dispersed nature necessitates the need to effectively coordinate and collaborate with people from different cultures, varying languages, incidence of different working practices and time zones variations (Chen et al. 2011; Bouchlaghem et al. 2005b). Given the incidence of virtuality in construction project teams over the past three decades, the fields of management, economy and technology are also constantly evolving to accommodate the concept of virtuality and its inherent challenges to teamwork, task function and interactions (Trautsch 2003; Manninen 2003).

Manninen (2003) and Hosseini et al. (2016) explored communicative and pragmatic aspects of interaction forms using a rich interaction model for Collaborative Virtual Environments (CVE) as a prelude to improving communication and interactions among project teams in the wake of increased virtuality. The application of decoding and encoding media richness through CVE is seen as a decisive response to the obvious lack of intuitive and non-intrusive non-verbal cues that separate computer-mediated communication settings from the usual synchronous face-to-face interactions among construction project teams (Manninen 2003; Hosseini et al. 2016). CVE-mediated communication platforms and models have the ability to enable project teams to take into account all the necessary manifestations and representations of interaction (Manninen 2003; Thalmann 2001; Allbeck and Badler 2001).

11.4 Theoretical Underpinnings of CVE and Rich Interactions in Project Teams

Various forms of interactions models extensively discussed and described in the existing literature in the field of communication have primarily focused face-to-face interaction between humans who share the same physical place with verbal and non-verbal communication channels and communication as main parameters (Argyle 1975; Manninen 2001, 2002; Laurel 1993; Benford et al. 1997). Within the context of technologically based on computer-supported collaborative work (CSCW), one area of focus has been interactions that relate to embodied actions that involve the movements and actions of the participants who interact with each other as well as their environment which is typical of virtual construction project environment (Manninen 2002; Laurel 1993; Robertson 1997).

It must be noted that rich interactions have both quantitative and qualitative attributes and measures that describe the amount and quality of available interaction forms (Manninen 2003; Laurel 1993). Manninen (2003) theorized that rich interactions can be facilitated by a set of interaction forms which is large, flexible and focused on the content, and thus by applying contextual and communicative support for interactions in collaborative virtual environment, it essentially provides users with meaningful ways to express themselves and their actions. By this, richness is achieved by users who are able to adapt and cautiously exploit the available interaction forms in an intuitive and non-deterministic style (Vilhjálmsson and Cassell 1998; Manninen 2002, 2003). Essentially, from a more pragmatic approach and synthesis of virtual team interactions and communication, the coverage has been on a wide area, consisting of CVE design tasks, communication and collaboration support (Manninen 2002, 2003).

The work of Benford et al. (1997) focused a spatial model of interaction which provided a basic set of abstractions for managing interactions in a wide range of spatial systems. This provided a novel approach to outlining powerful set of abstractions for managing interactions in a variety of large-scale virtual spaces that could bridge the virtual gap among the project teams involved in the interactions and communication. In Robertson (1997), an amalgamation of the taxonomy of embodied actions of the individuals and the physical workspace elements as well as others bodies formed the crux of the codes in the interactions. From the basis of these abstractions and arguments, the channel for effective CVE communication and interaction richness is theoretically attributed to the clarity in the definitions of the actions of the individuals and the physical workspace elements (Manninen 2002, 2003; Benford et al. 1997). In this, the aspect of non-verbal communications and interaction facets of CVEs in computer-supported collaborative working has extensively been focused on context of user embodiment, communicative behaviour, conversational interface agents and realistically expressing avatars (Benford et al. 1997; Cassell 2000; Thalmann 2001; Vilhjálmsson and Cassell 1998). However, it must be noted that the use of avatars for improving non-verbal cues and communication in construction CVEs is not extensively explored. However, drawing from Manninen (2003), when the focus of interactions and actions in CVE is on forms instead of functions, it can potentially enable user-driven communication, control and collaboration.

In CVE, relating typically of the construction environment, using 3D multiplayer action model to embed virtual communication platforms can impact on the effectiveness of the interactions, thus bridging the virtual divide (Manninen 2003). This can be significant given the digital age and heightened virtuality among project teams against the high volumes of project-related information, interactions and decisions required. The refinement provided by Manninen (2003) on CVE-rich interactions outlined autonomous, avatar appearance, chronemics, facial expressions, environmental details, kinesics, language-based communication, non-verbal audio, oculesics, olfactics, physical contact and spatial behaviour as the tenets of effective rich interaction models for CVEs. Hence, in construction project environment focusing on CVEs, operationalization of these tenets of effective rich interaction can engender communicative and pragmatic aspects of interaction forms among project teams in virtual communication. This will provide a novel innovation for analysing the CVE interactions and a design guideline in constructing new CVE experiments for construction project environment.

11.5 Potential Performance of CVE Interactions Models in Construction Project Teams

Generically, various interaction models in CVEs have proven potentials in enhancing interactions through non-verbal cues and communication in virtual environments (Benford et al. 1997; Cassell 2000; Thalmann 2001; Vilhjálmsson and Cassell 1998; Manninen 2003). In construction project environment, it can be said that learning from Manninen (2003), CVE interaction models and tools significantly influence related information in the interaction by structuring the data into coherent and descriptive categories. This attribute is essential for construction communication as this will lend support for sharing accurate project-related information and interaction leading to consistent and mutual understanding among the project team in the face of high-level virtuality.

In the case of ensuring a balance of synchronous and asynchronous communication among project teams as espoused by Otter and Emmitt (2007), CVE interaction models are deemed very helpful in pointing out several areas of interaction forms that were not adequately supported by the systems (Manninen 2003). This has become a panacea for developing critical knowledge in support of innovative virtual communication platforms that can bring about interaction richness in the area of information completeness, protocol consistency and access to information content. Gorse and Emmitt (2007) espoused that task-based and social-based interactions are common tenets to construction communication behaviours in group meetings and communications. In the notable literature, CVE interaction models draw theoretical comparisons with several approaches in social interactions and revealed a positive support for rich social interactions and activities (Manninen 2000, 2003).

An evaluation of CVE interaction model by Manninen (2003) focusing on process point of view on a value-added service production for mobile platforms for interactions showed that the significance of non-verbal communication in multiplayer game environments is theoretically demonstrated and practical which support rich interaction model and the equivalent design philosophy, thus providing an interaction model solution far superior developed purely with the technological focus. This according to Otter and Emmitt (2007) can bring about effective and prompt feedback to virtual communication loop and channels. The concept of rich interaction form model for CVEs illustrates various forms of communication, coordination and collaboration in virtual domains in construction project communication. Hence, an innovative and novel combination of various interaction forms, objects of the environment and embodied activities makes it possible to enhance the overall interaction and further increase the communicative, collaborative and constructive uses of the virtual environments (Manninen 2003; Cassell 2000; Thalmann 2001; Vilhjálmsson and Cassell 1998).

11.6 Behavioural Hindrances to Emerging Tools and Approaches to Information Sharing and Communication in Construction

Today's construction industry has witnessed an era of digitalization, increased virtuality and large volumes of project-related information and data that must be accurately shared among project teams and participants. From the perspective of Leidner (2010), these emerging characteristics of the industry today means, the issue of accurate information sharing grows proportionally with the progress of information and communication technology (ICT) adoption as enumerated in various industries. However, in spite of the upsurge of numerous tools and technology such as BIM, virtual reality and augmented reality as a platform for information sharing, issues from the context of information sharing behaviours still exist blighting the numerous gains that could have been realized from these ICT tools in the global construction industry.

Given the volume of information involved in construction project delivery alongside the incidence of digitalization and virtuality across various life cycle phases, there is an increasing necessity for adequate and efficient integration of information through appropriate ICT tools and platforms to facilitate positive project performance and information management (Egbu and Robinson 2005; Ibrahim et al. 2019). According to Otter and Emmitt (2007), project design team members from organizations using different information systems tend to have different understandings, opinions and rates of adoption and skill levels regarding specific IT tools. Navendren et al. (2014) enumerated professional role-related challenges to effectiveness of BIM in construction project delivery. From the extant literature, these are among other notable behavioural factors that hinder the optimization of the benefits of BIM and other ICT tools and platform for information sharing and documentation.

The incidence of BIM, VR and AR has generally been regarded as notable innovative 3D tools and platforms that are deemed to offer a revolutionizing way to design, document and procure buildings and information sharing and management on the basis of improved efficiencies and collaboration capabilities (Akponeware and Adamu 2017; Liu et al. 2017; Navendren et al. 2014). Significant behavioural factors and hindrances however exist which undermine the effectiveness and implementation of BIM, VR and AR platforms within the construction industry. Hence, the understanding of the behavioural factors that impact on the information sharing in BIM-, VR- and AR-mediated platforms has become an imperative precondition for their success, optimize their benefits across life cycle phases and facilitate process for the realization of the performance ambitions of the construction industry through integrated management of information in virtual 3D formats (Navendren et al. 2014; Singh et al. 2011; Becerik-Gerber et al. 2012). This understanding will also be significant of greatly aiding the improvement in the quality of information exchange among project teams (Singh et al. 2011; Garcia and Sayogo 2016; Becerik-Gerber et al. 2012; Hosseini et al. 2012, 2017).

This necessity is borne out of the practical fact that in spite of the surge of eagerness for ICT-mediated platforms (e.g. BIM, VR, AR, etc.), there is still the existence of persistent behavioural factors and issues impacting on information sharing among project teams within these platform-mediated projects. The possible reason contributing to this issue is attributed to behavioural factors and hindrances relating to lack of collaborative working approach, scarcity of knowledge among project teams and the existing culture of isolated working practices (Ibrahim et al. 2019; Eastman et al. 2008; Sacks et al. 2010; Garcia and Sayogo 2016; Merschbrock and Nordahl-Rolfsen 2016). Being successful either individually or organizationally in sharing project information is centrally dependent on awareness and understanding of organizational behaviours in sharing information (Cheng et al. 2013; Zhang and Ng 2012; Hosseini et al. 2012).

The growing urge for more effective and accurate ways of sharing and managing information among project teams in the construction industry means a more digital-based approach is required for real-time sharing of information to ensure integration, collaboration and effective management of the project through digital communication and information sharing (Ibrahim et al. 2019; Garcia and Sayogo 2016; Whyte and Donaldson 2015). This can again be effectively realized by a blend of effective information-sharing behaviours and context-specific ICT tools and platforms such as BIM, VR and AR. This is because a synthesis of both the social and technical aspects of ICT-mediated information platforms such as BIM and virtual reality will engender the needed outcome in the technical aspect of delivering project information (Ibrahim et al. 2019; Garcia and Sayogo 2016; Whyte and Donaldson 2015).

In understanding the behavioural factors that influence information sharing in ICT-mediated platforms such as BIM, VR and AR, from the extant literature, seven (7) key behavioural factors are deemed to significantly impact on information sharing among project teams in construction. These factors encompass trust, leadership, reciprocity, accountability, communication, culture and commitment (Zhang and Ng 2012; Wickramasinghe and Widyaratne 2012; Ho et al. 2013; Liu et al. 2017; Choi et al. 2008; Garcia and Sayogo 2016, etc.). The results from these studies lend credence to the fact that these behavioural factors have significant influence on the performance of information sharing among project teams (Javernick-Will 2012; Garcia and Sayogo 2016).

Trust-Related Factors El-Saboni et al. (2009): noted that trust has become a significant factor in ICT electronic communication systems in project delivery in United Arab Emirates, and this emanated from the experience of dishonesty and non-trustworthy from other parties fuelling reluctance on the part of others to freely share project information. Generally, when project team members exhibit dishonest and non-trustworthy behaviours, it tends to limit the zeal and willingness to share ideas, innovations and knowledge that are central to the progress of the project (Ibrahim et al. 2019; Wang et al. 2019). Ibrahim et al. (2019) found "truthfulness in dealing with others" as an important trust behavioural factor influencing successful information sharing among project teams in the Malaysian construction industry. It is said that lack of trust triggers uncertainties in sharing information, and thus, developing trust between participants is an effective means to reduce uncertainty in information sharing and communication among project teams (Wang et al. 2019; Strahorn et al. 2017). Wang et al. (2019) evaluated key issues in project team interactions and performance and found that the overall trust performance of the project team can instigate effective communication among project teams.

Reciprocity as a Behavioural Factor Wickramasinghe and Widyaratne (2012): outlined reciprocity as the next factor after trust that induces significant impact on information sharing in BIM-mediated projects. Javernick-Will (2012) indicated that reciprocation in information sharing is often associated with the feeling of indebtedness, gratitude and appreciation towards others where people often oblige to repay, to provide something in return or to treat others as they have been treated. Mutual respect is also often equated to reciprocation in information sharing and communication among group of participants (Ochieng and Price 2009; Diallo and Thuillier 2005). Ochieng and Price (2009) found mutual respect which is expressed in gratitude, honour respect for culture and ethics as an important factor for managing cross-cultural communication in multicultural construction project teams. In information sharing and communication in project-based organization, the behaviour of reciprocation enjoins communication interactants to repay others for what has been received from them or to treat others as they have been treated (Ibrahim et al. 2019; Javernick-Will 2012). From this, Javernick-Will (2012) intimated that generally in project based communication platforms, good intention and open attitude towards information sharing improvement has become a panacea for acceptable level of reciprocity.

Ibrahim et al. (2019) theoretically proved that, high incidence of reciprocity evidence in willingness to sharing information with the person that previously shared information with me as well as easily got help from others significantly impact on information sharing in BIM model communication and information mangement. This revelation suggests that reciprocity behavioural elements cannot be overlooked in ICT-mediated communication and information sharing among project teams.

Leadership Behavioural Factors the role of project leadership in information sharing and communication in project delivery has overly been cited as an important behavioural factor (see Garcia and Sayogo 2016; Wickramasinghe and Widyaratne 2012; Rogers et al. 2015; Navendren et al. 2014). It is theoretically asserted that project leadership remains one of the significant social aspects that affect the technical aspect of project activities especially in the area of information sharing (Navendren et al. 2014). There is evidence suggesting that sharing information among project teams breaks down when there is a lack of leadership in project delivery (Ibrahim et al. 2019; Cheng et al. 2013; Garcia and Sayogo 2016). Project team leadership in the context of communication often becomes the vehicle that drives openness, willingness, innovation, credibility and reliability of project information and exudes stability to the adopted tools for communication (Ahuja et al. 2010; Hosseini et al. 2017). Project leaders are so responsible for delineating the most suitable ICT protocols for sharing project-related information (Ahuja et al. 2009, 2010).

It is argued that project leaders must always serve as models by openly sharing information through the coordination of diverse viewpoints and also offer recognition for information shared (Garcia and Sayogo 2016). This motivates individuals to openly share information through the existing protocols as due recognition will be given (Garcia and Sayogo 2016; Cheng et al. 2013; Ho et al. 2013). The work of Ibrahim et al. (2019) on the Malaysian construction industry lends support to the fact that project leadership plays an important role in nurturing behaviour towards information sharing through encouragement from leaders to exchange information among team members in an open and transparent manner through the adopted protocols, tools and platforms.

Accountability Behavioural Factors lack of accountability in construction project delivery among project teams tends to derail team cohesion, increase tension and communication breakdown (Javernick-Will 2012; Ho et al. 2013; Latiffi et al. 2017). The measure of accountability in sharing information in project teams reflects in the ownership of information, responsibility of project leaders in sharing information, responsibility by all team participants in sharing information aimed at reducing project mistakes and discrepancies and a moral obligation to share information for the benefit of an organization task function (Ibrahim et al. 2019; Javernick-Will 2012; Ho et al. 2013; Latiffi et al. 2017). First and foremost, a good accountability practice must emanate from project leaders where they take charge and remain accountable for those they lead and at the same time hold others accountable (Ho et al. 2013; Latiffi et al. 2017).

According to Cheng et al. (2013) and Garcia and Sayogo (2016), the level of effectiveness in information sharing is tied to the accountability or moral obligation among the project teams. Thus, Ibrahim et al. (2019) alluded that project leaders assuming the full responsibility to share information emerged as the top behavioural element for accountability factor in BIM-based projects in Malaysia. It also equally found my moral obligation to share for the benefit of organization as being significant factor for information-sharing accountability. These suggest that project leaders must ensure open accountability to stimulate the right behaviours towards sharing of project information. This is significant towards improving communication and sharing information in BIM collaborative virtual environments.

Communication Behaviour Factors right communication behaviours are critical for effective information sharing no doubt. The literature is replete with the significance of communication to project success and process effectiveness (Ingason and Jónasson 2009; Jablin and Sias 2001). Poor communication behaviours can result in implicit communication performance that could further lead to other related work problems, thus making it extremely difficult to effectively engage team members in information sharing (Ho et al. 2013). Ibrahim et al. (2019) identified communication behaviour factors as the most important factor influencing information sharing in BIM projects. In Ibrahim et al. (2019), effective communication behaviours influencing information sharing take the form of prompt feedback from colleagues when required, easy to communicate with superiors and subordinates, communicating right information at the right time.

BIM by its nature and features is a communication-based technology for information sharing within a virtual environment facilitating the sharing of consistent information among the team (Mahamadu et al. 2013). Hence, right behaviours and skills through verbal and non-verbal means on BIM platforms allow for synchronizing information across applications that potentially speed up workflows and enable decision support, databases and purpose-driven content sharing (Redmond et al. 2012). Due to BIM, VR and AR being an integration and collaborative tools, it is imperative for project team participants to develop good communication behaviours in information sharing to ensure effectiveness in task function, understanding and access to the information.

Organizational Culture-Related Behaviour Factors The impact of culture on communication and information sharing is becoming an increasingly important area as this is also an important factor influencing the establishment of effective collaboration virtual environment communication platform especially in alliancing approach to projects (Deep et al. 2019; Bouchlaghem et al. 2005a; Ahuja et al. 2009). Ibrahim et al. (2019) opine that in project environment, the importance of understanding organizational culture in managing cultural diversity in generating information is emphasized due to the wide range of actors in construction projects.

Significantly, Ibrahim et al. (2019) deduced that a norm of frequently discussing progress of project openly is a positive cultural factor that impacts on information sharing among project teams. This behaviour breeds collectivism that is a positive feedback for the effectiveness of teams and organization (Xue et al. 2010; Navendren et al. 2014). Trigunarsyah (2017) inferred that organizational culture plays a major role in emphasizing teams' collective contribution and involvement in the construction project life cycle. This inference is significant in the fact that collective cultural behaviour places a shared obligation on project team members to everyone to share information as needed using the existing ICT protocols in BIM projects. This cultural behaviour also creates an environment that accommodates collaborative information sharing (Ho et al. 2013).

Good cultural behaviour factors in information sharing will result in helping each other whenever possible at all times, low difficulties between teams in sharing information with each other, active participation in information sharing inherent in positive organizational environment (Ibrahim et al. 2019; Hosseini et al. 2012).

Commitment Behaviour Factors Commitment of project teams to use adopted ICT protocols and platform for sharing project information is deemed as critical and central to success and project outcomes especially in the area of ensuring effective collaboration and integration of the team (Deep et al. 2019; Bouchlaghem et al. 2005a; Ahuja et al. 2009). Total commitment of the project team in sharing information is typified in working hard to ensure that the task given is completed and achieved the required objective and simply getting the work done as required (Choi et al. 2008; Ibrahim et al. 2019; Hosseini et al. 2012). Good commitment among project teams is required in moving towards a more partnering-like relationship for effective implementation of digital information sharing (Liu et al. 2017). Given that behaviour of sharing information cannot be forced but rather facilitated and stimulated, being able to facilitate good attitude in individuals and behaviour to share information consistently and willingly could help the daily operation of organizations (Liu et al. 2017; Huysman and de Wit 2002; Wei et al. 2012).

Inherent from the centrality of commitment to the behaviour of information sharing, being able to induce individuals to support and commit is crucially essential in initiating, leading and maintaining the high sense of cooperation within an organization that can significantly impact on information sharing through adopted ICT protocols (Ahuja et al. 2010). Ibrahim et al. (2019) alluded that collaborative working in multi-organizations without commitment in sharing the information derails the desire to help the organization to be successful by sacrificing self-interest, which in turn affects voluntary sharing of information affecting pro-social behaviour (Choi et al. 2008). Hence, stimulating commitment among project teams will facilitate pro-social behaviours towards effective information sharing.

11.7 Summary

The construction industry is at the crossroad of a paradigm shift from face-to-face synchronous means of communication to ICT-based asynchronous communication in the wake of an era of globalization and virtuality to project information and project teams, respectively. From this, various aspects of information sharing relating to technical dimension using ICT as well as the soft aspect have been pursued to be able realize this paradigm. In this chapter, a comprehensive overview has been given to highlight key ICT tools that can be applied to construction communication. Some of these tools were the Internet, intranet, project Website, groupwares, 3D and 4D collaborative virtual platforms. Additionally, the soft aspect aiding the performance of the technical dimension of communication tools has also highlighted trust, communication, accountability, leadership, organizational culture, reciprocity and commitment as key behavioural factors that influence ICT-based communication and information sharing in project teams. Against this, it can be said that the effectiveness of information sharing and communication in sharing in digital environment among construction project teams can be enhanced through a careful blend of both the technical and social aspects to create a better collective, integrative and collaborative information sharing in a virtual environment.

References

Ahmed, S., Hossain, M. M., & Hoque, M. I. (2017). A brief discussion on augmented reality and virtual reality in construction industry. *Journal of System and Management Sciences, 7*(3), 1–33.

Ahuja, V., Yang, J., & Shankar, R. (2009). Benefits of collaborative ICT adoption for building project management. *Construction Innovation: Information, Process, Management, 9*(3), 323–340.

Ahuja, V., Yang, J., & Shankar, R. (2010). IT-enhanced communication protocols for building project management. *Engineering, Construction and Architectural Management, 17*(2), 159–179.

Akponeware, A. O., & Adamu, Z. A. (2017). Clash detection or clash avoidance? An investigation into coordination problems in 3D BIM. *Buildings, 7*(3), 1–28.

Allbeck, J. M., & Badler, N. I. (2001). Consistent communication with control. Workshop on Non-verbal and Verbal Communicative Acts to Achieve Contextual Embodied Agents in Autonomous Agents, 2001.

Alshawi, M., Goulding, J., Khosrowshahi, F., Lou, E., & Underwood, J. (2009). *How strategic is IT investment in the construction industry?* (pp. 1–3). A UK Perspective, Modern Built Environment—Knowledge Transfer Networks, Intelligent Buildings Index, January.

Arayici, Y., Egbu, C. O., & Coates, P. (2012). Building information modelling (BIM) implementation and remote construction projects: Issues, challenges, and critiques. *Journal of Information Technology in Construction, 17*, 75–92.

Argyle, M. (1975). *Bodily communication.* New York: International Universities Press Inc.

Baldwin, A. N., Thorpe, A., & Carter, C. (1996). *The construction alliance and electronic information exchange: A symbiotic relationship.* CIB-65, Strathclyde University, Glasgow, UK.

Basu, A. (1996). *Impact of information technology on construction project management.* American Association of Cost Engineers Transactions, IT4.

Benford, S., Bowers, J., Fahlen, L. E., Greenhalgh, C., & Snowdown, D. (1997). Embodiments, avatars, clones and agents for multi-user, multi-sensory virtual worlds. *Multimedia Systems, 5*(2), 93–104.

Becerik-Gerber, A. A., Burcin, B., Ku, K., & Jazizadeh, F. (2012). BIM-enabled virtual and collaborative construction engineering and management. *Journal of Professional Issues in Engineering Education and Practice, 138*, 234–245.

Bouchlaghem, D., Shang, H., Anumba, C. J., Cen, M., Miles, J., & Taylor, M. (2005a). ICT-enabled collaborative working environment for concurrent conceptual design. *Architectural Engineering and Design Management, 1*(4), 261–280.

Bouchlaghem, D., Shang, H., Whyte, J., & Ganah, A. (2005b). Visualisation in architecture, engineering and construction (AEC). *Automation in Construction, 14*(3), 287–295.

Bråthen, K., & Moun, A. (2016). Bridging the gap: Bringing BIM to construction workers. *Engineering, Construction and Architectural Management, 23*(6), 751–764.

Brynjolfsson, E. (1993). The productivity paradox of information technology: Review and assessment. *Communications of the ACM, 36*, 66–78.

Brynjolfsson, E., & Hitt, L. M. (1998). Beyond the productivity paradox, computers are the catalyst for bigger changes. *Communications of the ACM, 41*(8), 49–55.

Cao, D., Li, H., Wang, G., Luo, X., Yang, X., & Tan, D. (2017). Dynamics of project-based collaborative networks for BIM implementation analysis based on stochastic actor-oriented models. *Journal of Management and Engineering, 33*, 04016055.

Cassell, J. (2000). Embodied conversational interface agents. *Communications of the ACM, 43*(4), 70–78; Ahuja, V., Yang, J., & Shankar, R. (2009). Study of ICT adoption for building project management in the Indian construction industry. *Automation in Construction, 18*, 415–423.

Chen, C.-H., Yang, J.-C., Shen, S., & Jeng, M.-C. (2007). A desktop virtual reality earth motion system in astronomy education. *Educational Technology & Society, 10*, 289–304.

Chen, F., Zhang, L., & Latimer, J. (2014). An exploratory study of social loafing in asynchronous virtual collaboration. *International Journal of Information and Decision Sciences, 6*(2), 153–165.

Chen, H., & Tien, H. C. (2007). Application of peer-to-peer network for real-time online collaborative computer aided design. *Journal of Computing in Civil Engineering, 21*(2), 112–121.

Chen, K., Lu, W., Peng, Y., Rowlinson, S., & Huang, G. Q. (2015). Bridging BIM and building: From a literature review to an integrated conceptual framework. *International Journal of Project Management, 33*(6), 1405–1416.

Chen, Y.-C., Chi, H.-L., Hung, W.-H., & Kang, S.-C. (2011). Use of tangible and augmented reality models in engineering graphics courses. *Journal of Professional Issues in Engineering Education and Practice, 137*, 267–276.

Cheng, L., Wang, Y., Hou, Z-G., Tan, M., & Cao, Z. (2013). Sampled-data based average consensus of second-order integral multi-agent systems: Switching topologies and communication noises. *Automatica, 49*(5), 1458–1464.

Choi, S. Y., Kang, Y. S., & Lee, H. (2008). The effects of socio-technical enablers on knowledge sharing: An exploratory examination. *Journal of Information Science, 34*(5), 742–754.

Dawood, N., Akinsola, A., & Hobbs, B. (2002). Development of automated communication of system for managing site information using internet technology. *Automation in Construction, 11*(3), 552–572.

Deep, S., Gajendran, T., & Jefferies, M. (2019). A systematic review of 'enablers of collaboration' among the participants in construction projects. *International Journal of Construction Management*, 1–13. https://doi.org/10.1080/15623599.2019.1596624.

Dehning, B., Dow, K. E., & Stratopoulos, T. (2004). Information technology and organizational slack. *International Journal of Accounting Information Systems, 5*, 51–63.

Diallo, A., & Thuillier, D. (2005). The success of international development projects: Trust and communication: An African perspective. *International Journal of Project Management, 3*(3), 237–252.

Duan, G., & Zhou, L. (2006). Study on product collaborative design system based on TeamCenter. *WSEAS Transactions on Computers, 5*(6), 1383–1388.

Duke, A., & Anumba, C. J. (1997). Enabling collaboration for virtual construction project teams. In C. J. Anumba & N. F. O. Evbuomwan (Eds.), *Concurrent engineering in construction* (pp. 163–172). London: Institution of Structural Engineers.

Eastman, C., Teicholz, P., Sacks, R., & Liston, K. (2008). *BIM handbook: A guide to building information modeling for owners, managers designers, engineers, and contractors*. Hoboken, NJ: Wiley.

Eastman, C. M., Teicholz, P., Sacks, R., & Liston, K. (2011). *BIM handbook: A guide to building information modeling for owners, managers, designers, engineers, and contractors*. New York, NY: Wiley.

Egbu, C. O., & Robinson, H. S. (2005). Construction as a knowledge-based industry. In C. Anumba, C. Egbu, & P. Carillo (Eds.), *Knowledge management construction* (pp. 31–49). Oxford: Blackwell Publishing.

El-Ghandour, W., & Al-Hussein, M. (2004). Survey of information technology applications in construction. *Construction Innovation, 3*, 83–98.

Ellis, C., & Wainer, J. (1994). A conceptual model of groupware. In R. Furutha & C. Neuwith (Eds.), *Proceedings of the ACM 1994 Conference on Computer Supported Cooperative Work* (pp. 79–88), Chapel Hill, October 22–26. ACM.

El-Saboni, M., Aouad, G., & Sabouni, A. (2009). Electronic communication systems effects on the success of construction projects in United Arab Emirates. *Advanced Engineering Informatics, 23*(1), 130–138.

Emmitt, S., & Gorse, C. A. (2007). *Communication in construction teams*. London: Taylor & Francis.

Garcia, J. R. G., & Sayogo, D. S. (2016). Government inter-organizational information sharing initiatives: Understanding the main determinants of success. *Government Information Quarterly, 33*(3), 572–582.

Golparvar-Fard, M., Peña-Mora, F., & Savarese, S. (2009). D4AR–a 4-dimensional augmented reality model for automating construction progress monitoring data collection, processing and communication. *Journal of Information Technology in Construction, 14*(13), 129–153.

Gorse, C. A., & Emmitt, S. (2007). Communication behaviour during management and design team meetings: A comparison of group interaction. *Construction Management and Economics, 25*, 1197–1213.

Graham, I., Spinardi, G., & Williams, R. (1996). Diversity in the emergence of electronic commerce. *Journal of Information Technology, 11*, 161–172.

Hardin, B., & Mccool, D. (2015). *BIM and construction management proven tools, methods and workflows* (2nd ed.). India: Wiley.

Ho, S. P., Tserng, H. P., & Jan, S. H. (2013). Enhancing knowledge sharing management using BIM technology in construction. *The Scientific World Journal, 2013*, 1–10.

Hosseini, M. R., Chileshe, N., Baroudi, B., Zuo, J., & Mills, A. (2016). Factors affecting perceived level of virtuality in hybrid construction project teams (HCPTs): A qualitative study. *Construction Innovation, 16*(4), 460–482.

Hosseini, M. R., Chileshe, N., Baroudi, B., Izadpanahi, P., & Zou, J. (2017). Virtuality of hybrid design teams within the construction context: A conceptual model. *International Journal of Project Organisation and Management, 9*(1), 1–17.

Hosseini, M. R., Chileshe, N., Zou, J., & Baroudi, B. (2012). Approaches for implementing ICT technologies within construction industry. *Australasian Journal of Construction Economics and Building: The Conference Series, 1*(2), 1–12.

Huysman, M. H., & de Wit, D. H. (2002). *Knowledge sharing in practice*. Dordrecht: Kluwer Academic Publisher.

Ibrahim, C. K. I. C., Sabri, N. A. M., Belayutham, S., & Mahamadu, A. (2019). Exploring behavioural factors for information sharing in BIM projects in the Malaysian construction industry. *Built Environment Project and Asset Management, 9*(1), 15–28.

Ingason, H. T., & Jónasson, H. I. (2009). Contemporary knowledge and skill requirements in project management. *Project Management Journal, 40*(2), 59–69.

Jablin, F. M., & Sias, P. M. (2001). Communication competence. In F. M. Jablin & L. L. Putnam (Eds.), *The new handbook of organizational communication: Advances in theory, research, and methods* (pp. 819–864). Thousand Oaks, CA: Sage.

Javernick-Will, A. (2012). Motivating knowledge sharing in engineering and construction organisations: Power of social motivations. *Journal of Management in Engineering, 28*(2), 193–202.

Jensen, P. A., & Jóhannesson, E. I. (2013). Building information modelling in Denmark and Iceland. *Engineering, Construction and Architectural Management, 20*(1), 99–110.

Kim, S.-Y. (2009). Modelling and analysis of a web-based collaborative information system—Petri net-based collaborative enterprise. *International Journal of Information and Decision Sciences, 1*(3), 238–264.

Latiffi, A. A., Brahim, J., & Fathi, M. S. (2017). Building information modelling (BIM) after ten years: Malaysian construction players' perception of BIM. In *2nd International Conference on Materials Science, Energy Technology and Environmental Engineering (MSETEE 2017)*, Zhuhai, April 28–30.

Laurel, B. (1993). *Computers as theatre*. Addison-Wesley Publishing Company, Inc.

Leidner, D. (2010). Globalization, culture and information: Towards global knowledge transparency. *Journal of Strategic Information Systems, 19*(1), 69–77.

Lin, K. Y., Son, J. W., & Rojas, E. M. (2011). A pilot study of a 3D game environment for construction safety education. *Journal of Information Technology in Construction (ITcon), 16*(5), 69–84.

Liu, Y. (2009). *Critical factors for managing project team communication at the construction stage* (Ph.D. thesis). Submitted to the Polytechnic University of Hong Kong.

Liu, Y., van Nederveen, S., & Hertogh, M. (2017). Understanding effects of BIM on collaborative design and construction: An empirical study in China. *International Journal of Project Management, 35*(4), 686–698.

Lou, M. M., Liew, P. S., Gu, N., & Ding, L. (2005). An agent approach to supporting collaborative design in 3D virtual worlds. *Automation in Construction, 14*(2), 189–195.

Love, P. E. D., Irani, Z., & Edwards, D. J. (2004). Industry-centric benchmarking of information technology benefits, costs and risks for small-to-medium sized enterprises in construction. *Automation in Construction, 13*(4), 507–524.

Love, P. E. D., Irani, Z., Li, H., & Cheng, E. (2001). An empirical analysis of IT/IS evaluation in construction. *Journal of Construction Information Technology, 8*(1), 15–27.

Mahamadu, A. M., Mahdjoubi, L., & Booth, C. A. (2013). Challenges to digital collaborative exchange for sustainable project delivery through building information modeling technologies. *WIT Transactions on Ecology and The Environment, 179*(1), 547–557.

Manninen, T. (2000). Interaction in networked virtual environments as communicative action—Social theory and multi-player games. In *Proceedings of CRIWG Conference* (pp. 154–157), Madeira, Portugal. IEEE Press.

Manninen, T. (2001). Rich interaction in the context of networked virtual environments—Experiences gained from the multi-player games domain. In *Proceedings of IHM-HCI* (pp. 383–398), Lille, France. Berlin: Springer.

Manninen, T. (2002). Interaction forms in multiplayer desktop virtual reality games. In S. Richir, P. Richard, & B. Taravel (Eds.), *Proceedings of VRIC2002 Conference*, 19–21 June, ISTIA Innovation (pp. 223–232).

Manninen, T. (2003). Conceptual, communicative and pragmatic aspects of interaction forms—Rich interaction model for collaborative virtual environments. In *Proceedings of Computer Animation and Social Agents (CASA) Conference* (pp. 168–174). IEEE Computer Society Press.

Marshall-Ponting, A. J., & Aouad, G. (2005). An nD modelling approach to improve communication process for construction. *Automation in Construction, 14*(3), 311–321.

Martinsons, M. G., & Martinsons, V. (2002). Rethinking the value of IT, again. *Communications of the ACM, 45*, 25–26.

Mead, S. P. (1997). Project specific intranets for construction project teams. *Project Management Journal, 28*(3), 44–51.

Mead, S. T. (1999). *Communication effectiveness in intranet based construction projects* (Unpublished Ph.D. thesis). Department of Civil and Building Engineering, Loughborough, University, Loughborough, UK.

Merschbrock, C., & Nordahl-Rolfsen, C. (2016). BIM technology acceptance among reinforcement workers—The case of Oslo airport's terminal 2. *Journal of Information Technology in Construction, 21*, 1–12.

Miettinen, R., & Paavola, S. (2014). Beyond the BIM utopia: Approaches to the development and implementation of building information modeling. *Automation in Construction, 43*, 84–91.

Mitchell, W. (1993). Don't wait for the data information super highway. *Construction Business Review*, 50–53.

Navendren, D., Manu, P., Mahamadu, A. M., Shelburn, M., & Wildin, K. (2014). Briefing: Towards exploring profession specific BIM challenges in the UK. *Proceedings of Institution of Civil Engineers: Management, Procurement and Law, 167*(4), 163–166.

Nitithamyong, P., & Skibniewski, M. J. (2004). Web-based construction project management systems: How to make them successful? *Automation in Construction, 13*(4), 491–506.

Nitithamyong, P., & Skibniewski, M. J. (2011). Success factors for the implementation of web-based construction project management systems. *Construction Innovation, 11*(1), 14–42.

O'Brien, M., & Al-Soufi, A. (1994). A survey of data communications in the construction industry. *Construction Management and Economics, 12*(5), 457–465.

Ochieng, E. G., & Price, A. D. (2009). Framework for managing multicultural project teams. *Engineering, Construction and Architectural Management, 16*(6), 527–543.

Otter, A., & Emmitt, S. (2007). Exploring effectiveness of team communication balancing synchronous and asynchronous communication in design teams. *Engineering, Construction and Architectural Management, 14*(5), 408–419.

Otter, A. F. H. J. (2005). *Design team communication using a project website* (Ph.D. thesis). Bouwstenen 98, Technische Universiteit Eindhoven, Eindhoven.

Park, C. S., Le, Q. T., Pedro, A., & Lim, C. R. (2015). Interactive building anatomy modeling for experiential building construction education. *Journal of Professional Issues in Engineering Education and Practice, 142*, 04015019.

Peansupap, V., & Walker, D. (2005). Factors affecting ICT diffusion: A case study of three large Australian construction contractors. *Engineering, Construction and Architectural Management, 12*(1), 21–37.

Perry, M., & Sanderson, D. (1998). Coordinating joint design work: The role of communication and artefacts. *Design Studies, 19*(3), 273–288.

Pittard, S., & Sell, P. (2016). *BIM and quantity surveying*. Abingdon: Routledge.

Rad, A. A., Ehsan Asnaashari, E., Knight, A., & Hurst, A. (2010). ICT utilization in administrative tasks in Iranian construction organizations. In C. Egbu (Ed.), *Proceedings 26th Annual ARCOM Conference* (pp. 615–624), September 6–8, 2010, Leeds, UK. Association of Researchers in Construction Management.

Redmond, A., Hore, A., Alshawi, M., & West, R. (2012). Exploring how information exchanges can be enhanced through Cloud BIM. *Automation in Construction, 24*, 175–183.

Robertson, T. (1997). Cooperative work and lived cognition: A taxonomy of embodied actions. In *European Conference on Computer Supported Cooperative Work* (pp. 205–220).

Rogers, J., Chong, H.-Y., & Preece, C. (2015). Adoption of building information modelling technology (BIM). *Engineering, Construction and Architectural Management, 22*(4), 424–445.

Rojas, E. M., & Songer, A. D. (1999). Web-centric systems: A new paradigm for collaboration engineering. *Journal of Management and Engineering, 15*(1), 39–45.

Root, D., & Thorpe, T. (2001). Refocusing collaboration technologies in construction supply chain: Looking beyond the organization. In *Proceedings of ARCOM, Seventeenth Annual Conference* (Vol. 1, pp. 253–262), University of Salford.
Sacks, R., Radosavljevic, M., & Barak, R. (2010). Requirements for building information modeling based lean production management systems for construction. *Automation in Construction, 19*(5), 641–655.
Sampaio, A. Z. (2018). Introduction of BIM methodology in education: Concept and application. In *4th International Conference on Higher Education Advances (HEAD'18)* (pp. 19–27), Editorial Universitat Politècnica de València.
Shen, L. Y. (1992). *Information management in construction companies*. Hong Kong: Hong Kong Polytechnic University, Department of Building and Surveying.
Singh, V., Gu, N., & Wang, X. (2011). A theoretical framework of a BIM-based multi-disciplinary collaboration platform. *Automation in Construction, 20*, 134–144.
Skibniewski, M. J., & Zavadskas, E. K. (2013). Technology development in construction: A continuum from distant past into the future. *Journal of Civil Engineering and Management, 19*(1), 136–147.
Stanney, K. M., & Hale, K. S. (2014). *Handbook of virtual environments: Design, implementation, and applications*. Boca Raton, FL, USA: CRC Press.
Strahorn, S., Brewer, G., & Gajendran, T. (2017). The influence of trust on project management practice within the construction industry. *Construction Economics and Building, 17*(1), 1–19.
Sutton, M. J. D. (1996). *Document management for the enterprise*. Chichester: Wiley.
Thalmann, D. (2001). The role of virtual humans in virtual environment technology and interfaces. In R. Earnshaw, R. Guejd, A. van Dam, & J. Vince (Eds.), *Frontiers of human-centred computing, online communities and virtual environments* (pp. 27–38). London: Springer.
Thorpe, T., Edum-Fotwe, F. T., & Mead, S. P. (1998). *Managing construction projects within emerging information driven business environments*. International Council for Building and Construction, Working Group W78, Stockholm.
Thorpe, T., & Mead, S. (2001). Project-specific websites: Friend or foe? *Journal of Construction Engineering and Management, 125*(5), 406–413.
Trautsch, B. R. (2003). *Managing virtual project teams* (MBA thesis). https://userwww.sfsu.edu/ceb/b895/btrautsch895.pdf. Accessed: June 21, 2019.
Trigunarsyah, B. (2017). Organizational culture influence on client involvement. *Engineering, Construction and Architectural Management, 24*(6), 1155–1169.
Vanier, D., & Turk, Z. (1995). Internet opportunities for distributed construction information. International Council for Building and Construction, Working Group W78, Information Technology in Construction. World Wide Web. https://delphLkstr.lth.se:80/w78/html.
Vetter, R. J., Spell, C., & Ward, C. (1994). Mosaic and the world wide web. *Computing Practices*, 49–57.
Vilhjálmsson, H. H., & Cassell, J. (1998). BodyChat: Autonomous communicative behaviors in avatars. In *Second International Conference on Autonomous Agents* (pp. 269–276), Minneapolis, MN, US.
Wang, D., Fu, H., & Fang, S. (2019). The efficacy of trust for the governance of uncertainty and opportunism in megaprojects: The moderating role of contractual control. *Engineering, Construction and Architectural Management*. https://doi.org/10.1108/ECAM-09-2018-0409.
Wang, X., & Love., P. E. D. (2012). BIM + AR: Onsite information sharing and communication via advanced visualization. In *Proceedings of the 2012 IEEE 16th International Conference on Computer Supported Cooperative Work in Design* (pp. 850–855). IEEE.
Wang, X., Truijens, M., Hou, L., Wang, Y., & Zhou, Y. (2014). Integrating augmented reality with building information modeling: Onsite construction process controlling for liquefied natural gas industry. *Automation in Construction, 40*, 96–105.
Wei, C., The, P., & Asmawi, A. (2012). Knowledge sharing practices. *Journal of Knowledge Management Practice, 1*(3), 111–123.

Whyte, A., & Donaldson, J. (2015). Digital model data distribution in civil engineering contracts. *Built Environment Project and Asset Management, 5*(3), 248–260.

Wickramasinghe, V., & Widyaratne, R. (2012). Effects of interpersonal trust, team leader support, rewards, and knowledge sharing mechanisms on knowledge sharing in project teams. *VINE, 42*(2), 214–236.

Wikforss, Ö., & Alexander, L. (2007). Rethinking communication in construction. *Journal of Information Technology in Construction, 12*, 337–346.

Wilson, P. (1991). *Computer supported cooperative work, an introduction to intellect*. New York: Oxford University Press.

Wong, A. K. D., & Zhang, R. (2013). Implementation of web-based construction project management system in China projects by Hong Kong developers. *Construction Innovation, 13*(1), 26–49.

Xie, C., Schimpf, C., Chao, J., Nourian, S., & Massicotte, J. (2018). Learning and teaching engineering design through modeling and simulation on a CAD platform. *Computer Applications in Engineering Education, 26*(4), 824–840.

Xie, X. (2002). *Communications in construction design* (Ph.D. thesis). Loughborough University, Loughborough.

Xie, C., Wu, D., Luo, J., & Hu, X. (2010). A case study of multi-team communications in construction design under supply chain partnering. *Supply Chain Management: An International Journal, 15*(5), 363–370.

Xue, X., Shen, Q., & Ren, Z. (2010). Critical review of collaborative working in construction projects: Business environment and human behaviors. *Journal of Management in Engineering, 26*(4), 196–208.

Yang, J., Ahuja, V., & Shankar, R. (2007). *Managing building projects through enchanted communication—An ICT based strategy for small and medium enterprises* (pp. 2344–2357). Paper presented at the CIB World Building Congress, Cape Town, South Africa, May, 2007.

Zhang, J., Wu, W., & Li, H. (2018). Enhancing building information modeling competency among civil engineering and management students with team-based learning. *Journal of Professional Issues in Engineering Education and Practice, 144*(2), 05018001.

Zhang, P., & Ng, F. F. (2012). Attitude toward knowledge sharing in construction teams. *Industrial Management & Data Systems, 112*(9), 1326–1347.

Part IV
Way Forward for Communicating in the Digital Era

Chapter 12
The Tenets for Improving Communication in Non-traditional Procurement

Abstract Several paradigms have been explored in the past as a sure way towards improving communication performance in the construction industry. It must indeed be admitted that though some of these paradigms are still relevant, others may not yield the needed impact in the wake of increasing complexities, dynamism, virtuality, internationalization of the industry, the structure of the project delivery process and diverse organizational structure as well as unique attributes of emerging preferred procurement typologies and management models. This limitations demands point to discover new directions and strategies that can suit these attributes and effect the needed improvement in communication, interactions and information sharing among project teams in project delivery. Here, we propose three main paradigms as key tenets to improving communication effectiveness and performance in non-traditional procurement systems. These entail ensuring digitilization of construction industry and organizations, dealing with virtuality and developing key communication competency behaviours. Though these paradigms are proposed with the focus on non-traditional procurement, they may equally be explored for conventional procurement models and management concepts. It is expected that by focusing on these a clear path that will provide the needed impetus in improving communication performance to engender effective project delivery.

Keywords Communication behaviours · Communication performance · Information sharing · Non-traditional procurement

12.1 Introduction

The urge to continuously improve communication performance in construction project delivery especially in non-traditional procurement has become an emerging necessity. The reflections provided in this book are a strong testament to the fact that, indeed, non-traditional procurement typologies experience different communication context, and thus, information sharing and interactions demand a new paradigm from the already established ones which suit conventional procurement systems. In this chapter, three main directions are proposed to engender improved communication performance in non-traditional procurement typologies given the current state

T. E. Kwofie et al., *Effective Construction Project Delivery*,
https://doi.org/10.1007/978-3-030-49374-5_12

of the industry. By considering digitalizations, dealing with virtuality and development of communication behaviours, it is expected that the needed improvement in communication performance can be actualized.

12.2 Understanding the Future of Construction Communication

The current construction industry has become intensely internationalized with notable diversified cultural orientation and dynamism that have significant implications on communication among project team. It is also well-known fact that the industry now is craving for a more collaborative working, improved relationship, increased integration and enhanced cooperation in the project delivery process. Additionally, construction industry is also at the stage of significant influence of the attributes of the emergence of the Fourth Industrial Revolution which is also known as the "Industry 4.0" (4IR). This characteristic of the current state of the industry calls for a rethink on the future of communication in the industry and project delivery process given its centrality to delivery success, performance and team effectiveness.

Interestingly in spite of these notable emerging characteristics of the industry and its implication on communication and information sharing, traditional models and strategies such as paper-based communication and information sharing are still dominating the industry. First, the notable increased virtuality of project team against the previous colocation of participants comes with it need for a communication paradigm that suits this attribute and at the same time enhances collaboration, integration, cooperation and improve relationship. Secondly, the obvious cultural diversities and dynamism from the internationalization of the industry from alliancing, partnership, joint venture and foreign construction organizations and team participants present obvious point of departure from existing communication paradigms and strategies to ones that can integrate and ameliorate the significant impact of these diversities. Similarly, another attribute of the industry which is inherent from the 4IR is the digitalization of information and the process for project delivery. In the wake of these, it is obvious to admit that the future of communication is certain to appoint of departure from current paradigm. The question is what should the industry do different to communicate effectively to overcome the threats and challenges posed by these attributes to communication.

A notable feature of the 4IR is the advancement in information and has seen every aspect of human society becoming more information oriented with the use of information technologies (Ozlu 2017). The situation in the construction industry is more pronounced. Though, the construction industry naturally saw the adoption of ICT tool as a precursor for improved information sharing, communication and task function (Otter and Emmitt 2007). However, with the obvious known development of individuals having different levels of understanding, opinions, skills and rates of adoption of the available communication tools as well as preferences for specific means of communication continues to pose a challenge to the effectiveness of available ICT tools to impact on communication outcome (Gorse 2002). Against this, the future of communication in the industry must be towards operationalizing ICT tools to overcome these challenges.

Obviously, the 4IR age is also known as the digital and technological age where there is extensive application of the combination of innovations and ICT tools such as the Internet of Things (IoT), sensors, cloud computing, robotics in conjunction with Artificial Intelligence (AI) among others for various aspects of production process including communication, information management and storage (Fuchs 2018). The application of these technological innovations through digital platforms to communication must be the future of construction communication. However, the implementation of the 4IR features is still in its naissance stages in the construction industry. Still construction organization and delivery process are dominantly human-based, non-digitized and lacking automation in spite of the benefits inherent from digitalization and automation. Additionally, Otter and Emmitt (2007) further stressed that effective communication among project teams is highly dependent on the communication acts of team members, their preferences for using specific communication media and access to easy to use tools and the competences of team interantants to facilitate, stimulate and motivate to communicate effectively as a team using the appropriate tools. These two factors lend credence to the suggestion of considering both technical and soft aspects of communication process towards improvement (Marshall-Pointing and Aouad 2005). From this, the argument that supports the future of communication in the construction industry must also give priority to developing key skills and competencies that can optimize the digital nature of the industry now.

12.3 What Should We Do Different to Enhance Effective Communication Outcome?

This is a question that various divergent but disjointed responses have been suggested and explored. Howbeit, this must be answered in one way or the other if significant progress and improvement in construction industry communication must be made that addresses the emerging digitalization age, growing virtuality and cultural and role diversities. Here, three main tenets are explored as a response to the above question.

12.3.1 Digitalization of Industry and Adoption of Digital Communication Platforms

Various scholars in built environment have stressed that the time is right to set up a real digital strategy for the construction industry. A real digital strategy has become an extremely essential for continuous success of the industry and improvement in project delivery process by facilitating communication, information sharing, access and storage. It is obvious that, as the pace of digital change is accelerating, the ability of firms in the construction industry to organize for digitalization has becoming increasingly imperative. Unfortunately, most construction organizations are yet to embrace this drive and become digitized. Digitalization of construction organization and adoption of digital platforms for communication and information sharing in construction project delivery will in no doubt facilitate a significant and momentous improvement in communication and access to information among project teams.

However, in spite of this recognition, the diffusion of diffusion of digital innovations in construction organizations and firms as a nonlinear process influenced by embedded contextual elements stemming from several complex social systems is still lacking and not extensively and rigorously pursued. Digitalizing construction organization and adoption of digital technology to processes and communication can produce radical innovations that alter existing industry architectures and nomenclatures by changing its power structures and organizational operating process (Morgan 2019). Cautiously, though the adoption of earlier digital tools and platforms exposed the industry to cybersecurity issues and risks, its impacts have been tremendous and positive and the accelerating rate of digitalization appears inevitable (Shaw et al. 2017; Morgan 2019).

Existing digitalization technological innovations such as BIM have been noted to have improved communication, information sharing, easy access to information and understanding information through 3D and 4D virtual models of drawings and texts (Davies and Harty 2012). Notably, a more digitalized industry and construction organizations as well as embracing automation will tend to create a digital value chain to facilitate the conveyance of information in the products and their environment and business partners (Oesterreich and Teuteberg 2016; Vaidya et al. 2018). The industry can also explore the integration of automation and robotics to the value chain which will also help improve information sharing and communication in the delivery process. In no doubt, this digitalization drive will facilitate effective collaboration, integration, cooperation, improve relationships and cooperation which is desperately needed in the industry to optimize the benefits of non-traditional procurement typologies which has these attributes.

12.3.2 Dealing with Virtuality in Construction Communication

The construction industry is extensively exposed to various forms of virtuality in the delivery process which have significant impact on task function, social interactions and information sharing. Thus, a more collaborative working environment making use of computer-supported cooperative and collaborative working (CSCW) model is imperative to overcoming the challenges inherent from virtuality of the project delivery environment. This will facilitate a form of virtual collaborative communication, information-sharing environment and task and social-based interaction forms. By facilitating communication through collaborative virtual environment will tend to influence many aspects of organization and management which can induce the needed effectiveness in the delivery process of construction projects.

Indeed Lou et al. (2005) provided practical insights affirming that the use of multi-agent system to support collaborative design in 3D virtual worlds are able to facilitate a level of communication not readily available in conventional computer-aided design (CAD) systems. The benefit of the collaborative virtual communication environment is its ability to ensure coherency of product information in the design process through integrating CAD system with product data management (PDM) system something missing in conventional tools and communication model. Hence, an urge towards a more collaborative working environment using CSCW will be indeed a welcome approach to bridge the virtuality gap and influence communication outcome among the virtual project teams.

12.3.3 Developing Communication Competency Behaviours

Construction project teams are predominantly multi-disciplinary and communication among the team is deemed to be a collective function (Otter and Emmitt 2007). Thus, effective collective communication among multi-disciplinary project teams is highly dependent on the ability and willingness of all the team members to act and react, to listen and share as well as develop their skills for using communication effectively (Forsyth 2006). From the perspective of Otter and Emmitt (2007), multi-disciplinary project teams need both the means and skills communicate effectively among the team. This is an indication that development of key communication skills and competencies can be critical as well as serving as an approach through which these key skills can be legitimized towards effective communication outcomes.

However, the notable setback of the industry is a situation where all other skills to perform have been rigorously pursued at the neglect of communication skills. By comparison, other industries and sectors regarding the necessity of communication have made adoption and development of communication skills central to their training and education. This is a sharp contract to the construction sector. There is yet to be any profession in the built environment that embraces communication and communication competencies as central tenet of the training process or in continuous professional development. This trend needs to change. Otter and Emmitt (2007) further intimated that one key aspect of ensuring effective collective communication among multi-disciplinary project teams is the ability of teams to posses the needed skills to use communication media, channels, protocols and tools collectively. From this insight, it is clear that the future of construction communication must embrace development of communication competencies as an imperative.

Through these issues of compatibility of adopted electronic means of communication and tools as well as differences of opinion and understanding on an individual level, including differences in the use of specific electronic means for team communication can be bridged. This will affirm the soft aspect component towards communication improvement in project delivery.

12.4 Summary

Here, in this chapter, a point of departure from existing conventional models of communication in construction to emerging tenets that facilitate the current digital state, increased virtuality and competency behaviours of the industry is espoused. In no doubt, embracing digitalization of the industry and organization, developing communication competencies and bridging the virtual divide through computer-supported collaborative virtual communication platforms is the tenets for communication in the industry now. By this, the needed improvements in communication among the project team towards delivery effectiveness.

References

Davies, R., & Harty, C. (2012). Control, surveillance and the "dark side" of BIM. In S. D. Smith (Ed.), *Proceedings of the 28th Annual ARCOM Conference*, Edinburgh, UK. London, UK: Association of Researchers in Construction Management.

Forsyth, D. R. (2006). *Group dynamics* (4th ed.). Belmont, CA: Thomson Wadsworth.

Fuchs, C. (2018). Industry 4.0: The digital German ideology. *Triple C (Cognition, Communication, Cooperation): Open Access Journal for a Global Sustainable Information Society, 16*(1), 280–289.

Gorse, C. (2002). *Effective interpersonal communication and group interaction during construction management and design team meetings* (Ph.D. thesis). University of Leicester, Leicester, UK.

Lou, M. M., Liew, P. S., Gu, N., & Ding, L. (2005). An agent approach to supporting collaborative design in 3D virtual worlds. *Automation in Construction, 14*(2), 189–195.

Marshall-Ponting, A. J., & Aouad, G. (2005). An nD modelling approach to improve communication process for construction. *Automation in Construction, 14*(3), 311–321.

Morgan, B. (2019). Organizing for digitalization through mutual constitution: The case of a design firm. *Construction Management and Economics, 37*(7), 400–417.

Oesterreich, T. D., & Teuteberg, F. (2016). Understanding the implications of digitisation and automation in the context of Industry 4.0: A triangulation approach and elements of a research agenda for the construction industry, 121–139.

Otter, A., & Emmitt, S. (2007). Exploring effectiveness of team communication balancing synchronous and asynchronous communication in design teams. *Engineering, Construction and Architectural Management, 14*(5), 408–419.

Ozlu, F. (2017). The advent of Turke's industry 4.0. *Turkish Policy Quarterly, 16*(2), 29–38.

Shaw, J. D., Bansal, P., & Gruber, M. (2017). New ways of seeing: Elaboration on a theme. *Academy of Management Journal, 60*(2), 397–401.

Vaidya, S., Ambad, P., & Bhosle, S. (2018). Industry 4.0, a glimpse. *Procedia Manufacturing, 20*(11–12), 233–238.

The manufacturer's authorised representative in the EU is Springer Nature Customer Service Centre GmbH, Europaplatz 3, 69115 Heidelberg, Germany. If you have any concerns regarding our products, please contact ProductSafety@springernature.com

Printed and bound by CPI Group (UK) Ltd, Croydon, CR0 4YY
30/09/2026
02225697-0005